CO_2 Capture and Renewable Energy

CO_2 Capture and Renewable Energy

Editors

Marta G. Plaza
Rui P. P. L. Ribeiro

MDPI • Basel • Beijing • Wuhan • Barcelona • Belgrade • Manchester • Tokyo • Cluj • Tianjin

Editors

Marta G. Plaza
Instituto de Ciencia y
Tecnologia del Carbono
(INCAR), CSIC
Spain

Rui P. P. L. Ribeiro
Universidade NOVA de
Lisboa
Portugal

Editorial Office
MDPI
St. Alban-Anlage 66
4052 Basel, Switzerland

This is a reprint of articles from the Special Issue published online in the open access journal *Energies* (ISSN 1996-1073) (available at: https://www.mdpi.com/journal/energies/special_issues/CO2_Capture_Renewable_Energy).

For citation purposes, cite each article independently as indicated on the article page online and as indicated below:

LastName, A.A.; LastName, B.B.; LastName, C.C. Article Title. *Journal Name* **Year**, *Volume Number*, Page Range.

ISBN 978-3-0365-4959-0 (Hbk)
ISBN 978-3-0365-4960-6 (PDF)

Cover image courtesy of Marta G. Plaza and Rui P. P. L. Ribeiro

Contents

About the Editors vii

Preface to "CO_2 Capture and Renewable Energy" ix

Marta G. Plaza and Rui P. P. L. Ribeiro
Special Issue "CO_2 Capture and Renewable Energy"
Reprinted from: *Energies* **2022**, *15*, 5187, doi:10.3390/en15145187 1

Maria João Regufe, Ana Pereira, Alexandre F. P. Ferreira, Ana Mafalda Ribeiro and Alírio E. Rodrigues
Current Developments of Carbon Capture Storage and/or Utilization–Looking for Net-Zero Emissions Defined in the Paris Agreement
Reprinted from: *Energies* **2021**, *14*, 2406, doi:10.3390/en14092406 5

Marta G. Plaza, Sergio Martínez and Fernando Rubiera
CO_2 Capture, Use, and Storage in the Cement Industry: State of the Art and Expectations
Reprinted from: *Energies* **2020**, *13*, 5692, doi:10.3390/en13215692 31

Wahiba Yaïci, Evgueniy Entchev and Michela Longo
Recent Advances in Small-Scale Carbon Capture Systems for Micro-Combined Heat and Power Applications
Reprinted from: *Energies* **2022**, *15*, 2938, doi:10.3390/en15082938 59

Jarosław Krzywanski, Waqar Muhammad Ashraf, Tomasz Czakiert, Marcin Sosnowski, Karolina Grabowska, Anna Zylka, Anna Kulakowska, Dorian Skrobek, Sandra Mistal and Yunfei Gao
CO_2 Capture by Virgin Ivy Plants Growing Up on the External Covers of Houses as a Rapid Complementary Route to Achieve Global GHG Reduction Targets
Reprinted from: *Energies* **2022**, *15*, 1683, doi:10.3390/en15051683 89

Rui P. P. L. Ribeiro, Isabel A. A. C. Esteves and José P. B. Mota
Adsorption of Carbon Dioxide, Methane, and Nitrogen on Zn(dcpa) Metal-Organic Framework
Reprinted from: *Energies* **2021**, *14*, 5598, doi:10.3390/en14185598 97

Ana Almeida, Rui P. P. L. Ribeiro, José P. B. Mota and Carlos Grande
Extrusion and Characterization of High Si/Al Ratio ZSM-5 Using Silica Binder
Reprinted from: *Energies* **2020**, *13*, 1201, doi:10.3390/en13051201 113

About the Editors

Marta G. Plaza

Marta González Plaza is a Tenured Scientist at Instituto de Ciencia y Tecnología del Carbono (INCAR), from the Agencia Estatal Consejo Superior de Investigaciones Científicas (CSIC). She completed her MSc in Chemical Engineering at the Universities of Oviedo, Spain, and Strathclyde, UK, in 2004. In 2009, she got her PhD in Chemical Engineering from the University of Oviedo, Spain, after completing two research stays in the University of Nottingham, UK. She worked as a post-doctoral researcher in the Laboratory of Separation and Reaction Engineering (LSRE), at the University of Porto, Portugal, between 2009 and 2011, and then moved to INCAR, in Spain. Marta G. Plaza has been awarded with several fellowships throughout her career, including the Ramon y Cajal programme. She is the author of over 50 scientific articles, the inventor of one patent, and has participated in over 20 research projects. Her main research interests are CO_2 capture and bioenergy.

Rui P. P. L. Ribeiro

Rui Pedro Pinto Lopes Ribeiro is a Researcher at LAQV@Requimte Associate Laboratory at the NOVA School of Science and Technology (Portugal). He has a PhD in Chemical and Biological Engineering from the University of Porto (Portugal), obtained in 2013. Rui Ribeiro has published over 30 peer-reviewed papers and contributed to more than 60 communications in scientific meetings. Rui Ribeiro's main research interests include the design and optimization of separation processes (e.g., CO_2 capture, biogas upgrading, etc.), and the characterization and formulation of novel adsorbent structures via extrusion and 3D printing.

Preface to "CO_2 Capture and Renewable Energy"

This book provides a collection of six peer-reviewed articles that cover a range of subjects and applications related to CO_2 Capture and Renewable Energy. These hot topics are of interest not only to specialists and researchers working in the field, but also to those working in industries and those in the general public who wish to learn about the status and perspectives of these key enablers against climate change.

The first section of the book is composed of three review articles: the first provides a general overview of CO_2 capture and storage, while the second focuses on CO_2 capture in the hard-to-abate cement sector, and the third concerns CO_2 capture in small-scale heat and power applications.

The second section of the book is a collection of three novel research articles: Chapter 4 evaluates the mitigation potential of implanting ivy plants on the roofs and walls of buildings worldwide, whereas Chapters 5 and 6 deal with the development of solid adsorbents for CO_2 capture and the application of biogas upgrading.

Marta G. Plaza and Rui P. P. L. Ribeiro
Editors

energies

Editorial

Special Issue "CO_2 Capture and Renewable Energy"

Marta G. Plaza [1,*] and Rui P. P. L. Ribeiro [2,*]

1 Instituto de Ciencia y Tecnología del Carbono (INCAR), CSIC, C/Francisco Pintado Fe 26, 33011 Oviedo, Spain
2 LAQV-REQUIMTE, Department of Chemistry, NOVA School of Science and Technology, NOVA University of Lisbon, 2829-516 Caparica, Portugal
* Correspondence: m.g.plaza@incar.csic.es (M.G.P.); rpp.ribeiro@fct.unl.pt (R.P.P.L.R.)

This book contains the successful submissions [1–6] to the Special Issue of *Energies* on the wide subject area of "CO_2 Capture and Renewable Energy".

The urgently-needed carbon neutral economy can only be accomplished through a portfolio of strategies, among which CO_2 capture and renewable energy will need to play a decisive role.

The widespread adoption of renewable energy is an unquestionable part of the solution against climate change. Even though hydropower has traditionally led renewable electricity generation, wind and solar are growing fast. These technologies have rapidly evolved from niche to mainstream thanks to policy drivers, technology developments and international co-operation that have reduced their cost, making them the cheapest source of electricity in many markets. However, dispatchable renewables, such as hydropower, bioenergy, concentration solar power and geothermal, will still be critical to maintain electricity security, together with other low-carbon generation, energy storage and robust electricity networks. Renewables will also play a significant role to produce heat in the industry and the residential sector, while reducing the emissions from the transport sector. Biomethane, produced by biogas upgrading, is also known as renewable natural gas. This can be directly blended with, or fully replace natural gas in existing pipelines and end-user equipment, with the added advantage of being carbon neutral.

In the transition to a sustainable energy economy, based on renewables and green energy sources, CO_2 capture and storage (CCS) will be of paramount importance for abating CO_2 emissions from existing infrastructure in the power and the industrial sectors. Furthermore, there are many industries which are very hard or impossible to decarbonise in the short term, such as the cement sector, in which CO_2 emission are intrinsic to the production process. In such cases, CCS will be mandatory to achieve the goal of net zero emissions. Alternatively, the captured CO_2 can be utilised (CCU) in various applications. The latter approach is receiving more attention as it provides an added value to the captured CO_2. However, it is important to highlight that if the origin of the CO_2 is fossil, only CCS or long-term CCU, such as enhanced oil recovery or construction materials, will contribute to combat climate change. Shorter-lived CCU, such as chemicals and synthetic fuels, can only provide neutral emissions pathways in a circular economy when making use of CO_2 from biogenic origin (biomass), or of CO_2 previously removed from the atmosphere by technological means (known as direct air capture: DAC). Permanent CO_2 removal technologies, such as bioenergy with carbon capture and storage (BECCS) and direct air capture and storage (DACS) are foreseen as necessary in the medium term to compensate for emissions from the hard-to-abate sectors, and in the long term, even to remove atmospheric CO_2 from past emissions.

In this Special Issue, an interesting collection of works covering the topics referred above are presented. The publications include three Research Articles and three Review Articles.

Almeida et al. [1] report the preparation of ZSM-5 extrudates with potential for adsorption-based biogas upgrading. The extrudates were prepared using a silica-based

Citation: Plaza, M.G.; Ribeiro, R.P.P.L. Special Issue "CO_2 Capture and Renewable Energy". *Energies* **2022**, *15*, 5187. https://doi.org/10.3390/en15145187

Received: 11 July 2022
Accepted: 14 July 2022
Published: 18 July 2022

binder, and the adsorption equilibrium of CO_2, CH_4 and H_2O was determined. The extrudates have mechanical resistance at least comparable to commercial zeolites, a surface area reduction lower than 10% when compared to the pristine powders, and similar CO_2/CH_4 selectivity.

Plaza et al. [2] review the progress of CO_2 capture technologies already evaluated in the cement industry at pilot scale, and the established plans for near-future commercial demonstration. The authors highlight the necessity of CCUS to abate at least the 65% of emissions intrinsically related to the cement production process (calcination of limestone). Five large scale projects are on the pipeline to demonstrate CO_2 capture in the cement sector, and these involve different technologies, such as amine absorption, adsorption, calcium looping, and direct capture through indirect heating for calcination.

The developments in the wider area of CCUS are reviewed by Regufe et al. [3]. The authors frame the need for CCUS with the historic evolution and the latest data regarding CO_2 emissions. Then, they briefly review the different types of CO_2 capture technologies and their technology readiness level. This is followed by a general overview of the current progress of CCUS facilities and their barriers to deployment, with special focus on BECCS and DAC. Finally, the status, cost, and prospects of CCUS in industrial processes are also briefly overviewed.

Ribeiro et al. [4] evaluate the potential of Zn(dcpa) metal-organic framework (MOF) for application in CO_2 capture and biogas upgrading. The adsorption equilibrium of CO_2, CH_4, and N_2 is reported. An interesting finding is highlighted: the isotherm of CO_2 at 273 K has a stepwise shape with hysteresis assumed to be related with two metastable structures of the MOF. The authors provide an interpretation of this behaviour in terms of the osmotic thermodynamic theory.

The possibility of implanting virgin ivy plants on house walls and roofs to fix atmospheric CO_2 is proposed by Krzywanski and co-workers [5]. According to these authors, this approach should be considered among the climate change mitigation portfolio, as it could lead to an annual CO_2 removal of over 3.5 Gt (*ca.* 7% of global greenhouse gas emissions), while releasing oxygen, reducing dust in the environment, and lowering the air temperature.

Wahiba et al. [6] review the latest advances in small-scale carbon capture systems, with special focus in decentralized, small-scale combined heat and power (CHP) cogeneration facilities, including micro gas turbines (mGT) coupled to post-combustion amine absorption, hybrid solid state fuel cell systems coupled to mGT and chemical looping for H_2 generation, and biomass-fired organic Rankine cycle (ORC) coupled to post-combustion amine absorption. The main challenges of integrating amine scrubbing post-combustion carbon capture into mGT are identified as the high volumetric flow rate of the exhaust gas, its low CO_2 concentration (1.5 vol%), and its O_2 content that leads to solvent degradation, which results in a high energy penalty. Exhaust gas recirculation (EGR) reduces the amount of exhaust gas fed to the capture unit and increases its CO_2 concentration, thus reducing the energy penalty and the capital costs. The EGR ratio is limited by an optimal O_2 concentration at the combustor inlet, but CO_2 can be further enriched making use of S-EGR. Humidification is found to increase mGT cycle efficiency, compensating energy losses from EGR. The authors highlight that hybrid fuel cell systems coupled to mGT and chemical looping for H_2 generation have the major benefit of 100% capture, while ORC-based biomass-fired micro-CHP integrated with post-combustion capture can offer net atmospheric removal of CO_2. Carbon capture is found to be the costliest phase of CCUS supply chain, specially at small-scale applications. Truck and railroad tankers are identified as a lower cost option for CO_2 transport in small scale applications compared to large-scale pipelines.

In sum, we believe this collection of articles is an important contribution to the open literature in the areas of CO_2 Capture and Renewable Energy.

Author Contributions: The authors contributed equally to this work. All authors have read and agreed to the published version of the manuscript.

Funding: Rui Ribeiro acknowledges financial support from FCT/MCTES through the Norma Transitória DL 57/2016 Program Contract.

Acknowledgments: M.G.P. and R.P.P.L.R. are grateful to the authors that contributed to this Special Issue, and to MDPI for the invitation to act as guest editors and for the support provided throughout the whole publishing process.

Conflicts of Interest: The authors declare no conflict of interest.

References

1. Almeida, A.; Ribeiro, R.P.P.L.; Mota, J.P.B.; Grande, C. Extrusion and Characterization of High Si/Al Ratio ZSM-5 Using Silica Binder. *Energies* **2020**, *13*, 1201. [CrossRef]
2. Plaza, M.G.; Martínez, S.; Rubiera, F. CO_2 Capture, Use, and Storage in the Cement Industry: State of the Art and Expectations. *Energies* **2020**, *13*, 5692. [CrossRef]
3. Regufe, M.J.; Pereira, A.; Ferreira, A.F.P.; Ribeiro, A.M.; Rodrigues, A.E. Current Developments of Carbon Capture Storage and/or Utilization–Looking for Net-Zero Emissions Defined in the Paris Agreement. *Energies* **2021**, *14*, 2406. [CrossRef]
4. Ribeiro, R.P.P.L.; Esteves, I.A.A.C.; Mota, J.P.B. Adsorption of Carbon Dioxide, Methane, and Nitrogen on Zn(dcpa) Metal-Organic Framework. *Energies* **2021**, *14*, 5598. [CrossRef]
5. Krzywanski, J.; Ashraf, W.M.; Czakiert, T.; Sosnowski, M.; Grabowska, K.; Zylka, A.; Kulakowska, A.; Skrobek, D.; Mistal, S.; Gao, Y. CO_2 Capture by Virgin Ivy Plants Growing Up on the External Covers of Houses as a Rapid Complementary Route to Achieve Global GHG Reduction Targets. *Energies* **2022**, *15*, 1683. [CrossRef]
6. Yaïci, W.; Entchev, E.; Longo, M. Recent Advances in Small-Scale Carbon Capture Systems for Micro-Combined Heat and Power Applications. *Energies* **2022**, *15*, 2938. [CrossRef]

energies

MDPI

Review

Current Developments of Carbon Capture Storage and/or Utilization–Looking for Net-Zero Emissions Defined in the Paris Agreement

Maria João Regufe *, Ana Pereira, Alexandre F. P. Ferreira, Ana Mafalda Ribeiro and Alírio E. Rodrigues

LSRE-LCM-Laboratory of Separation and Reaction Engineering-Laboratory of Catalysis and Materials, Associate Laboratory, Faculdade de Engenharia, Universidade do Porto, Rua Dr. Roberto Frias, 4200-465 Porto, Portugal; ajmp@fe.up.pt (A.P.); aferreir@fe.up.pt (A.F.P.F.); apeixoto@fe.up.pt (A.M.R.); arodrig@fe.up.pt (A.E.R.)

* Correspondence: mjregufe@fe.up.pt

Abstract: An essential line of worldwide research towards a sustainable energy future is the materials and processes for carbon dioxide capture and storage. Energy from fossil fuels combustion always generates carbon dioxide, leading to a considerable environmental concern with the values of CO_2 produced in the world. The increase in emissions leads to a significant challenge in reducing the quantity of this gas in the atmosphere. Many research areas are involved solving this problem, such as process engineering, materials science, chemistry, waste management, and politics and public engagement. To decrease this problem, green and efficient solutions have been extensively studied, such as Carbon Capture Utilization and Storage (CCUS) processes. In 2015, the Paris Agreement was established, wherein the global temperature increase limit of 1.5 °C above pre-industrial levels was defined as maximum. To achieve this goal, a global balance between anthropogenic emissions and capture of greenhouse gases in the second half of the 21st century is imperative, i.e., net-zero emissions. Several projects and strategies have been implemented in the existing systems and facilities for greenhouse gas reduction, and new processes have been studied. This review starts with the current data of CO_2 emissions to understand the need for drastic reduction. After that, the study reviews the recent progress of CCUS facilities and the implementation of climate-positive solutions, such as Bioenergy with Carbon Capture and Storage and Direct Air Capture. Future changes in industrial processes are also discussed.

Keywords: CO_2 emissions; CCS; CCUS; global facilities

Citation: Regufe, M.J.; Pereira, A.; Ferreira, A.F.P.; Ribeiro, A.M.; Rodrigues, A.E. Current Developments of Carbon Capture Storage and/or Utilization–Looking for Net-Zero Emissions Defined in the Paris Agreement. *Energies* **2021**, *14*, 2406. https://doi.org/10.3390/en14092406

Academic Editor: Attilio Converti

Received: 22 March 2021
Accepted: 9 April 2021
Published: 23 April 2021

1. Introduction

Due to the Industrial Revolution, fossil fuels (coal, oil and gas) were unlocked as a new energy resource. Their excessive use has led to several negative impacts such as global climate change [1].

Figure 1 presents the surface air temperature anomalies between 1979 and 2020. For example, comparing with September 2016, September 2019 was the warmest month in the data record, which was 0.57 °C warmer than the average temperature from 1979–2010 [1].

Natural causes are also responsible for this effect. However, greenhouse gases from human activities including population growth, deforestation, agriculture, urbanization (urban heat islands), and the resulting changes in consumption patterns are responsible for more than 95% of global warming [2]. The greenhouse contribution in the temperature variation is evident, responsible for about 1 °C, more than other anthropogenic sources and natural variations.

The principal greenhouse gas related to global climate change is carbon dioxide (CO_2). Other gases, in minor quantities, are also responsible, such as methane (CH_4), water vapor (H_2O), nitrous oxide (N_2O) and fluorinated gases (F-gases), specially hydrofluorocarbons (HFCs), perfluorocarbons (PFCs), sulfur hexafluoride (SF_6), and nitrogen trifluoride (NF_3) [3,4].

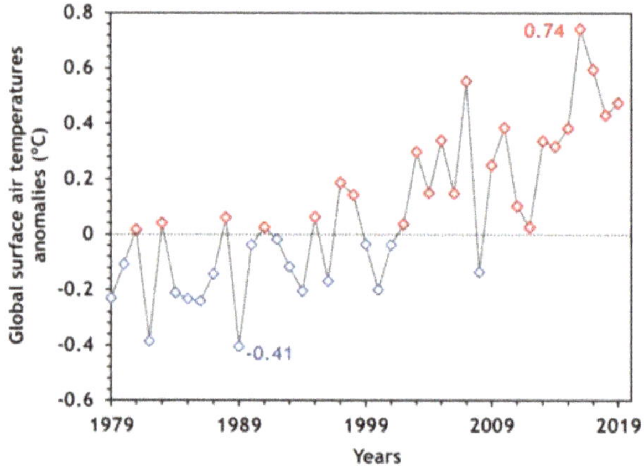

Figure 1. Global mean surface air temperature anomalies in the 1979–2020 period (adapted from Service [1]).

In order to combat climate change and achieve decarbonization, several methods have been studied and implemented: a massive development of clean energies (renewable energy sources); fossil fuel consumption reduction by switching to lower-carbon alternatives, e.g., coal to gas; energy efficiency increase in industrial applications and the power sector, particularly in technologies used to convert fossil fuels into energy; carbon capture and utilization/storage techniques [2,5–7]. The employment of all of the options mentioned above will be required because CO_2 emission abatement became a global priority. However, at the current state of development, the levels of risks and the costs, non-fossil fuel energy alternatives cannot meet our need for energy fed by fossil fuels. Additionally, any quick change to non-fossil energy sources, even if this action was possible, would result in large disruptions to the existing energy supply infrastructure with substantial consequences to the global economy [8].

In the Paris Agreement, 196 parties decided to establish a long-term goal to keep the worldwide average temperature increase below 2 °C above pre-industrial levels and limit the increase to 1.5 °C, since this would significantly reduce the risks and effects of climate change [9–11]. Consequently, to obtain a sustainable low carbon future, global CO_2 levels should be drastically reduced by promoting the actions and investments needed. This limitation in the worldwide temperature implies immediate and decisive actions on climate change to avoid some of the worst climate impacts and reduce the chances of extreme weather occurrences around the world.

Thus, to meet mid to long-term CO_2 emissions targets, cost-effective CO_2 capture from fossil fuel use and subsequent sequestration options need to be evaluated, as well as the utilization of (captured) carbon dioxide as a feedstock for new products. Figure 2 shows the CO_2 life-cycle considering two pathways: carbon capture and storage (CCS) and carbon capture and utilization (CCU).

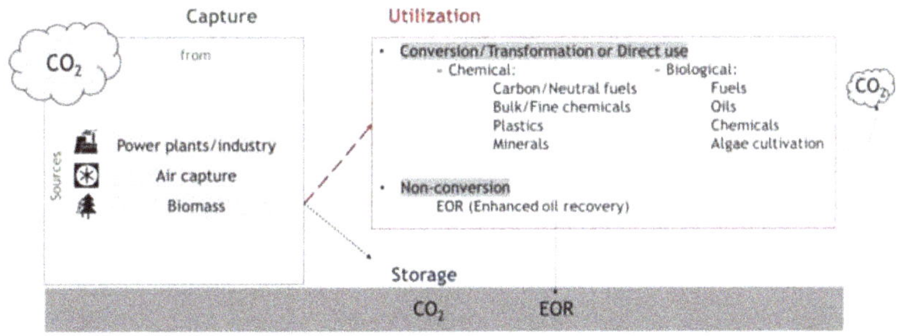

Figure 2. Analysis of life-cycle of CO_2 with capture and storage, and capture and utilization from main sources pathways (based on Sekera and Lichtenberger [12] and Bui et al. [13]).

Broadly recognized as having an enormous potential to meet climate change targets, CCS and CCUS appear as solutions to deliver low carbon heat and power, decarbonize the industry, and, more recently, facilitate the net removal of CO_2 from the atmosphere [14].

This article aims to review the overall CCS and CCUS strategies implemented to fulfil the climate change ambition established in the Paris Agreement of 2015. The approach to obtain a climate-neutral–an economy with net-zero greenhouse gas emissions-implies large changes in all the economic sectors, as well as energy, transport, industry, and agriculture.

First, the values of actual CO_2 emissions are presented, and it is analyzed the impact of COVID-19 in the first quarter of 2020. After that, Carbon Capture and Storage and Carbon Capture, Utilization, and Storage strategies to combat CO_2 emissions are described. CO_2 capture technologies, classified into three groups, precombustion, oxy-fuel and post-combustion systems, are presented as well as the leading technologies used for CO_2 capture. CCUS's current development, focusing on the facilities and projects working in Europe, is presented.

Two commonly used technologies of climate-positive solutions are briefly described, which are bioenergy with carbon capture and storage (BECCS) and direct air capture (DAC). In conclusion, the future of industrial processes, that use fossil fuels as raw materials and release CO_2 emissions, is analyzed.

2. CO_2 Emissions

In 2019, global energy-related CO_2 emissions reached 33 gigatonnes (Gt), approximately [15]. This resulted mainly from a sharp decline in CO_2 emissions from the power sector in advanced economies (Australia, Canada, Chile, European Union, Iceland, Israel, Japan, Korea, Mexico, Norway, New Zealand, Switzerland, Turkey, and United States.), because of the expanding role of renewable sources (mainly wind and solar photovoltaic systems), fuel switching from coal to natural gas, and higher nuclear power output. However, the total emissions, in the rest of the world, increased.

Figure 3 shows the gigatonnes of CO_2 emitted by developed countries, the rest of the world, and total emissions from 1990 until 2019 [16].

Figure 4 shows the global greenhouse gas emissions (%) by sector in 2020 [16]. The economic sector which had the highest share of carbon dioxide emissions from fossil fuels and cement was the power sector. With a 44% of emissions, this was more than the combined share of both industry and surface transport. These three sectors of the economy make up the majority of the world's CO_2 emissions.

Covid-19 had an enormous impact on energy demand and, therefore, on CO_2 emissions. The drastic curtailment of global economic activity and mobility during the first quarter of 2020 pushed down global energy demand by about 3.8% compared with the first quarter of 2019 [15]. CO_2 emissions were about 5% lower in Q1 2020 than in Q1 2019, almost twice as large as all previous declines since the end of World War II. By sectors, emissions from coal, oil, and natural gas declined about 8, 4.5, and 2.3%, respectively. By regions, a considerable decrease of CO_2 emissions was observed: −8% in China, −8% in the European Union (EU), and −9% in the United States [15]. However, this decline was punctual; it will not be enough to resolve climate change problems.

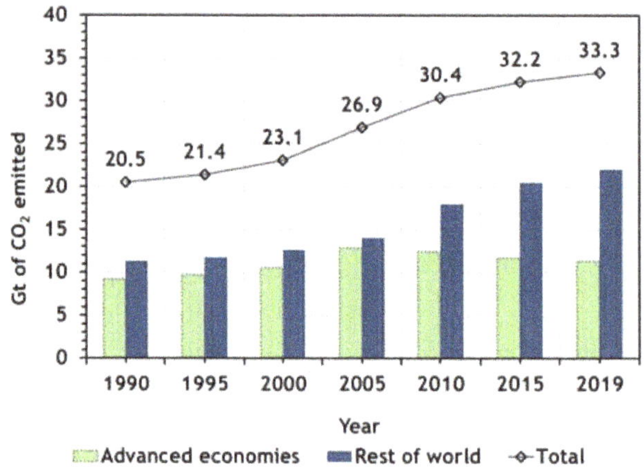

Figure 3. CO_2 emissions by countries for a period from 1990 until 2019 (adapted from Agency [16]).

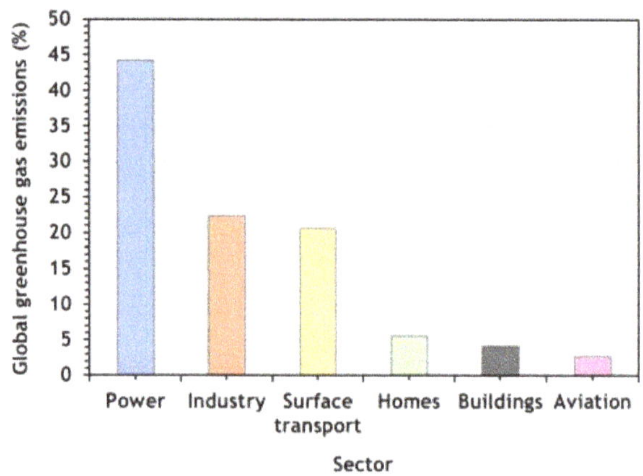

Figure 4. Global greenhouse gas emissions by sector in 2020 (adapted from Tiseo [17]).

In April 2020, McKinsey Global Institute published an overview of several scenarios of projected global CO_2 emissions that helps to understand the future. Climate Action Tracker [18] also provide these data. The scenarios are shown in Figure 5, where global CO_2 emissions in each scenario are projected. All pathways include energy-related emission, industry-process emissions (e.g., from cement production), emissions from deforestation and waste, and negative emissions (e.g., from reforestation and carbon-removal technologies such as bioenergy with carbon capture and storage, and direct air carbon capture and storage). Emissions from biotic feedbacks (e.g., from permafrost thawing, wildfires) were not considered. The red lines represent warming projections if policies are not applied: the lower bound is a "continued growth" pathway based on the IEA's World Energy Outlook 2019 current policies scenario; the higher bound is based on IPCC's Representative Concentration Pathway 8.5 [19]. It is possible to observe the need for immediate reduction of GHG emissions.

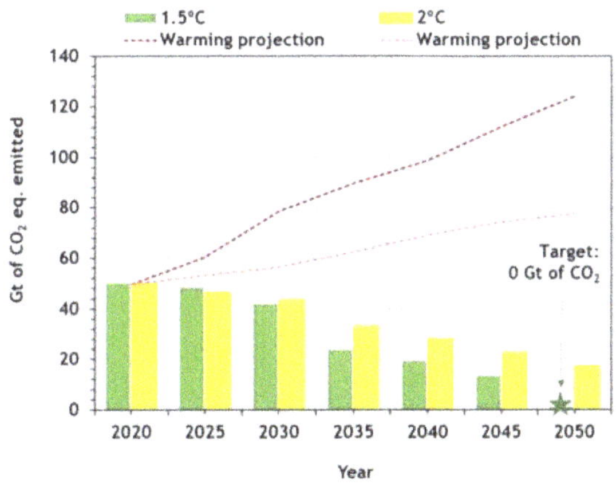

Figure 5. Projected global CO_2 emissions per scenario (adapted from Climate Action Tracker [18]).

According to this study, if no changes are applied, the continued growth will lead to about 120 Gt equivalent of CO_2 emitted per year in 2050. However, to achieve the 1.5 °C pathway of the Paris Agreement, the CO_2 emissions should be 0 Gt by then.

3. Carbon Capture (Utilization) and Storage (CCUS or CCS)

Basically, carbon capture and storage (CCS) consists of the separation and concentration of CO_2 from power generation plants or industrial processes, its pressurization and transportation, via ship or pipeline, to specific locations where it should be permanently stored deep underground, in geological formations (depleted oil or gas reservoirs or deep saline aquifers) [20,21]. This technology has been identified as a priority, being a critical emissions reduction technology that can be applied across the energy system, expecting to play an essential role in meeting the global warming targets [22–24].

CCS is often used interchangeably with the term Carbon Capture, Utilization, and Storage (CCUS). The difference between the two terms presented is the 'utilization' word, which refers to the use of carbon for other applications. CCUS can contribute to almost one-fifth of the emissions reductions needed across the industry sector. CCUS will play a key role in reducing CO_2 emissions from fossil-fuel-based power generation and is the only option available to reduce direct emissions from other industrial point sources significantly [25]. It was estimated that the use of CCUS would address up to 32% of global CO_2 emissions reduction by 2050 [26]. More than 28 Gt of CO_2 could be captured from industrial processes until 2060, the majority of it from the cement, steel, and chemical subsectors [27].

CCS and CCUS technologies are developed slowly, mainly as a result of high costs and unsupportive policy and regulatory frameworks in many countries [28].

The economic penalty of the capture is the crucial obstacle to CCS/CCUS implementation. The efficiency of the CO_2 capture must be increased in the capture step of the processes, as it is estimated that the capture step is responsible for 60% to 80% of the overall CCS/CCUS economic penalty [8,20]. The capture part of the process represents the main promise for cost reduction and focuses on most of the research efforts. CCS or CCUS is far from the ideal solution because it does not directly use green fuels. Still, it is the only technology capable of maintaining the utilization of the existing power plants.

In these types of processes, CO_2 capture technologies can be classified into three groups: pre-combustion systems, post-combustion systems, and oxy-fuel or oxy-combustion processes. The first and second systems depend on whether carbon dioxide is removed before or after fuel is burned. In the third, pure oxygen rather than air is used for combustion [3]. Figure 6 shows a brief scheme of methods for carbon capture.

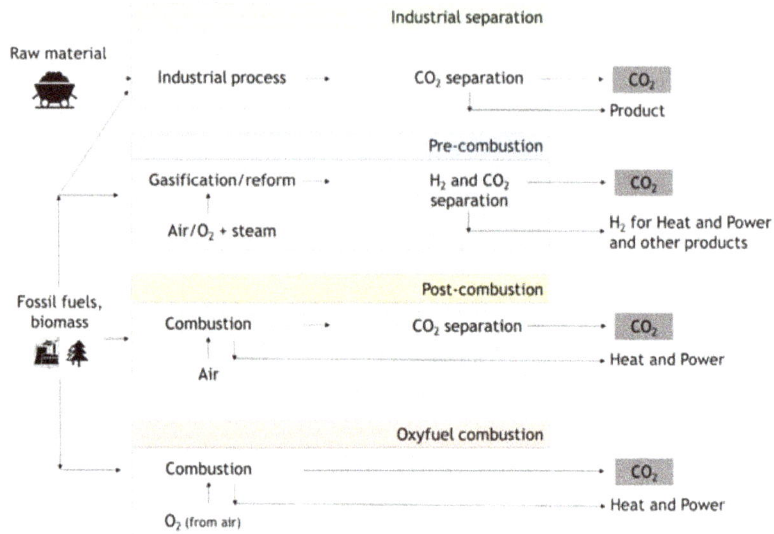

Figure 6. Scheme of methods for carbon capture (adapted from IPCC [29]).

In the production processes, namely in activities related to CO_2 and other harmful greenhouse gases, even in coal power plants, it is important to reduce the emissions. For that, energy production changes as technology advances. Energy companies and industries use several technologies. This way, Industry 4.0 and the external environment force the energy companies to constantly adjust goals [2,30]. Oxy-combustion capture is still under development and is not yet commercial. Reduction of NOx, SOx, Hg emissions, and methods of exhaust gas dedusting are also important.

3.1. Pre-Combustion

In power plants, in oil, gas, and chemical industries processes, where CO_2 is produced, the pre-combustion CO_2 capture can be used [3]. Technologies that separate this gas from gas streams have been used for many decades. The main objective of the industries is CO_2 removal to meet the required downstream product specifications, whether natural gas, hydrogen, or chemicals.

In pre-combustion CO_2 capture systems, the fuel source is decarbonized before combustion. More recently, in anticipation of the requirements to limit CO_2 emissions, plants design have been improved to convert the gas produced from gasification to hydrogen and CO_2 and remove CO_2 before the combustion of the hydrogen-rich gas in the turbine [31]. The gasification or partial oxidation process combines the reacting coal with steam and oxygen at high pressure and temperature. The product is a gaseous fuel consisting mainly of carbon monoxide and hydrogen, called synthesis gas or syngas.

After this, syngas is converted to more hydrogen and carbon dioxide by adding steam at a lower temperature. This is the Water Gas Shift Reaction (WGS) (Equation (1)). Before the combustion of the hydrogen-rich gas in the gas turbine, the CO_2 is captured. The concentration can be in the range of 15–60% (dry basis/% volume), and the total pressure is typically 2–7 MPa [3,31,32]:

$$CO + O \leftrightarrow CO_2 + H_2 \tag{1}$$

$$\Delta H = -40.6\,\text{kJ mol}^{-1}$$

The WGS reaction is the desired route for industrial applications, most commonly in conjunction with the Fischer-Tropsch (FT) reaction to synthesize hydrocarbon fuels from syngas. The conditions used for the FT reaction lie in the range of 200–375 °C; lower temperatures for long-chain alkanes and higher temperatures for shorter [32].

When compared with post-combustion process, CO_2 presents a higher concentration in the pre-combustion gas stream (>20% in the H_2 + CO_2 stream vs. 5–15% in a post-combustion flue gas stream). Then, CO_2/H_2 separation is somewhat more straightforward than the CO_2/N_2 separation in the post-combustion process due to the difference in molecular weights and molecular kinetic diameters [33].

CO_2 and H_2 can be separated using several technologies. Solvent-based CO_2 capture can be applied by chemical or physical (such as the Selexol and Fluor processes) absorption of CO_2 from syngas into a liquid carrier and regenerating the absorption liquid by increasing the temperature or reducing the pressure to break the absorbent-CO_2 bond [34]. Sorbent, membrane, and hybrid systems that combine attributes from multiple technologies are under investigation to reduce costs and energy penalties, as well as, to improve performance [35].

3.2. Oxy-Combustion

The oxy-combustion processes were designed to remove the bulk nitrogen from the air before combustion. A combination of oxygen (95% of purity, approximately) and recycled flue gas is used for the fuel combustion. A mixture with CO_2 and H_2O is generated by recycling the flue gas, and this mixture is ready for sequestration without stripping of the CO_2 from the gas stream [36]. The flame temperature is controlled by the amount of recycled flue gas. No chemical solvent or physical sorbent is required to separate CO_2 from the flue gas due to the high concentration in the stream. The carbon dioxide rich flue gas would then be delivered by pipeline to be sequestered.

This system was developed as an alternative to the more conventional post-combustion process in coal-fired power plants. The main reason is the reduced cost of oxy-combustion when compared with post-combustion. However, although good results were obtained in laboratory scale and pilot plants, commercial plants use is still scarce [3].

3.3. Post-Combustion

Post-combustion CO_2 capture systems have been used for many decades, and in this process, the CO_2 is captured from the products of burning fossil fuels (coal, natural gas, or oil) or combustion exhaust gases. The flue gas passes through a liquid solvent, solid adsorbent, membrane, or another medium, depending on the method/technology, allowing the separation of the CO_2 from the mixture. After that, CO_2 can be transported and stored.

The drawback of post-combustion carbon capture is the low carbon dioxide concentration in the flue gases, which leads to a relatively high energy penalty and high costs of carbon capture. On the other hand, pre-combustion strives to reduce these penalties by decarbonizing the process stream before combustion, resulting in more favorable conditions and more flexible implementation, significantly reducing capture costs [37].

Several technologies can be applied for separating or capturing CO_2 from a mixture of gases in an industrial process. The purification step and the technical approach used depend on the gas stream conditions, such as temperature, pressure, and concentration, and on the product purity required.

The captured and purified gas will be transported to its final destination. In the case of CCS, a pipeline is necessary to transport captured CO_2 for a storage site. When CCU is applied, a spur on the pipeline can take a slipstream from the main flow to be diverted to the chemicals or synthetic fuels plant. At the end of the supply chain, a minor quantity of CO_2 could still be emitted or stored [38].

3.4. Technologies for CO_2 Capture

Figure 7 shows technical approaches available for CO_2 separation and capture.

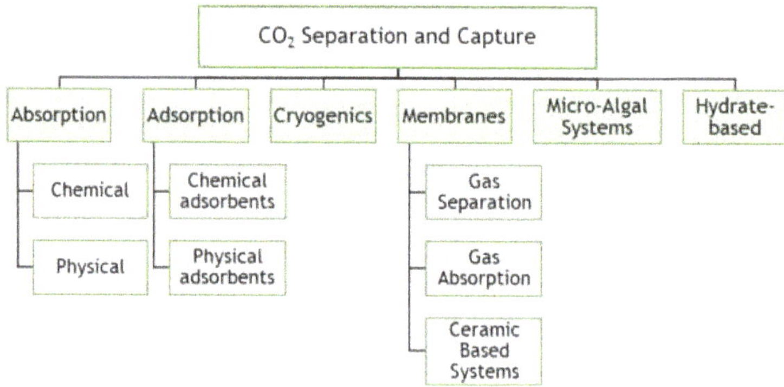

Figure 7. Technical options for CO_2 capture processes (adapted from Songolzadeh, Ravanchi and Soleimani [21]).

The most common process used to separate the CO_2 from natural gas, refinery off-gases, and synthesis gas processing is absorption technology [39]. This is characterized by using a liquid/solvent that selectively absorbs CO_2 from a gas stream. Afterwards, the solvent can be regenerated through a stripping or regenerative process by heating and/or pressurization [40]. Absorption processes can be chemical absorption, used in pre-combustion or post-combustion capture, or physical absorption, primarily used in pre-combustion capture. Selexol (with dimethyl ethers of polyethylene glycol solvent), Rectisol (with methanol solvent), and Purisol (with N-methyl-2-pyrolidone as solvent) are the most common physical processes. Typical chemical solvents are primary amines such as monoethanolamine (MEA) and 2-amino-2-methyl-1-propanol (AMP), secondary amines such as diethanolamine (DEA), and ternary amines such as methyldiethanolamine (MDEA) [39,41]. However, in this type of process, gas streams are required at high pressure. Plants for CO_2 capture with processes based on chemical absorption using MEA solvent were developed over 75 years ago to remove acidic gas impurities like H_2S and CO_2 from natural gas streams. Afterwards, the process was adapted to treat flue gas streams, and with this technology, about 85 to 95% of the CO_2 is captured, and a product stream of CO_2 can be produced with a purity higher than 99% [42].

The major challenges for CO_2 capture from flue gases by absorption processes are the sizeable volumetric flow rates at atmospheric pressure with large amounts of CO_2 at low partial pressures (10–15% of CO_2) at 40 °C. Then, the process presents several disadvantages, which are the high energy consumption due to the high thermal energy required, around 4.0 GJ/t of CO_2 captured [41] (considering 30 wt% MEA and 90% CO_2 removal), the presence of SOx and NOx contaminants, and the high oxygen partial pressure, which hinders the implementation of amine absorption process [43]. Besides, it leads to corrosive product formation due to the solvents' thermal and oxidative solvent degradation. There are many studies about processual alternatives to reduce the costs involved in power plants to reduce the operating costs. Besides the physical and chemical absorption methods discussed above, other methods could be implemented, as verified in Figure 7.

Gas separation through adsorption processes can be used in pre- and post-combustion capture and are promising alternative separation techniques characterized by solid adsorbents capable of reversibly capturing CO_2. Novel adsorbent materials for CO_2 capture with specific properties can adsorb large amounts of CO_2 to be used or stored, being these materials instruments for CO_2 utilization and storage. Adsorbents are porous solids and have a large surface area per unit mass. Each type of molecule or component creates different interactions with the adsorbent surface, leading to an eventual separation [44].

There are many types of adsorbents, which could be applied to CO_2 capture by physical adsorption processes, including activated carbons, carbon fibers, zeolites [45], metal-organic frameworks [46], and organic-inorganic hybrid materials [47,48]. The adsorbent should be chosen taking into account economic and operational criteria, which are (i) high adsorption capacity for the target gas component, i.e., CO_2, leading to the reduction of the adsorbent quantity and process equipment size; (ii) high CO_2 selectivity, representing a high adsorption capacity ratio between CO_2 and the other components in the stream, such as, nitrogen; (iii) fast adsorption and desorption kinetics; (iv) good physical and chemical stability during the cycles and regeneration steps; (v) be regenerable by modest pressure decrease or temperature increase, leading to the minimization of the operating energy costs. Furthermore, the adsorbent should ideally also have robust performance in the presence of moisture and other contaminants that may be present in the gas stream to treat. Then, there are essential features that should be considered for a successful operation of adsorbent material, such as composition, particle size, pore size, and pore connectivity.

Depending on the regeneration method, adsorption processes can be denominated as pressure swing adsorption (PSA), temperature swing adsorption (TSA), and electrical swing adsorption (ESA) [49,50].

Cryogenic carbon capture utilizes the principle of separation based on the cooling of CO_2 to low temperature. The CO_2 is separated from the flue gas mixture after cooling this gas below -73.3 °C at atmospheric pressure. After this, CO_2 is pressurized and delivered at pipeline pressure. Cryogenic separation can be applied for post-combustion processes in two different ways. In one of these methods, CO_2 can be de-sublimated to solid CO_2 on the heat exchangers, further heated and pressurized to obtain liquid CO_2 in the recovery stage. Clodic and Younes [51] proposed this type of separation. Tuinier et al. [52] proposed another method, with the use of packed beds for de-sublimation of CO_2. CO_2 is recovered from the packing material by feeding a fresh gas stream to increase the temperature and enhance the concentration of the CO_2 recovered from the packed bed [53]. It may be a good technique because it does not involve any additional chemicals in the separation process. However, the high compression power requirements for this method are the major disadvantage [54].

Membranes are another potential alternative to conventional solvent absorption technology. The difference in physical and/or chemical interactions between gases and membrane materials is responsible for the CO_2 separation. The method presents many advantages, such as reduced equipment size, lower energy requirements, simplicity in the process, among others. Nevertheless, in the post-combustion process, particularly in the CO_2/N_2 separation, due to the relatively low CO_2 concentration and pressure, the driving force for membranes to perform appropriately is weak, making their implementation difficult [55].

Another potential technique for removing CO_2 from flue gases is microalgae. Microalgae are microscopic organisms that typically grow suspended in water and are driven by the same photosynthetic process as higher plants [56].

Microalgal cells are sunlight-driven cell factories that can convert carbon dioxide into raw materials for producing biofuels (e.g., biohydrogen, biodiesel, and bioethanol), animal food chemical feedstocks, and high-value bioactive compounds [56].

The ability of these cells to absorb CO_2 can be applied as an attractive alternative for CO_2 sequestration. CO_2 fixation and storage via microalgae are essentially photosynthesis, transforming water and CO_2 into organic compounds without extra energy addition or consumption and secondary pollution.

Hydrate-based CO_2 capture (HBCC) technology emerges as a potential solution for CO_2 capture from gas streaming, e.g., fromCO_2/N_2 or from CO_2/H_2 of fossil fuel power plants. This technology is based on the hydrate cages formation by water molecules at high pressure and low temperature, where CO_2 molecules stay enclathrated, allowing their separation. It is estimated that this technology could have a cost reduction of CO_2 capture of about 45% when compared with the chemical absorption technology [57]. Recently, studies involving hydrate-base CO_2 capture and storage have increased [58,59].

3.5. Current Progress of CCUS Facilities

Since 1972, CCS has been applied to capture CO_2 from an extensive range of sectors and industries [7]. Typically, the progress of technology development contains a series of scale-up steps: first, laboratory scale or bench; second, pilot-scale; third, demonstration-scale; fourth, commercial scale. Currently, there are eighteen large-scale facilities in operation in the world, five under construction, and twenty in various stages of development [14] (see Table 1).

Table 1. Current development progress of technologies in terms of technology readiness level (TRL): carbon capture; transport; storage; and utilization (adapted from Bui [13],Consoli [14]).

Technology Readiness Level	**Current Development**
TRL1	Concept
TRL2	Formulation
S	Ocean Storage
TRL3	Proof of concept (lab tests)
C	Ionic Liquids-Post-combustion
C	BECCS Power
C	Low T separation-Pre-combustion
C	Membranes dense inorganic (CO_2 separation)
S	Mineral storage
TRL4	Lab prototype
C	Oxy-combustion gas turbine (water cycle)
TRL5	Lab-scale plant
C	Membranes dense inorganic (H_2 separation for reformer)
TRL6	Pilot plant
C	Membranes polymeric (power plants)
C	Biphasic solvents-Post-combustion
C	Chemical looping combustion (CLC)
C	Calcium carbonate looping (CaL)
U	CO_2 utilization (non-EOR)
TRL7	Demonstration
C	Membranes polymeric (NG industry)
C	Pre-combustion IGCC + CCS
C	Oxy-combustion coal power plant
C	Adsorption-Post-combustion
C	BECCS industry
C	DAC
S	Depleted oil & gas fields
S	CO_2-EGR

Table 1. *Cont.*

Technology Readiness Level	Current Development
TRL8	Commercial Refinement required
TRL9	Commercial
C	Post-combustion amines (power plants)
C	Pre-combustion NG processing
C	Transport on-shore & off-shore pipelines
T	Transport ships
S	Saline formations
S	CCUS

Notes: BECCS corresponds to bioenergy with CCS, IGCC corresponds to integrated gasification combined cycle, EGR corresponds to enhanced gas recovery, EOR corresponds to enhanced oil recovery, NG corresponds to natural gas; CO_2 utilization (non-EOR) reflects a wide range of technologies, most of which have been demonstrated conceptually at the lab scale. C–Capture; T–Transport; U–Utilization; S–Storage.

3.6. Global Facilities of CCUS

More than 30 new integrated CCUS facilities have been announced since 2007, mostly in the United States and Europe, although projects are also planned in China, Australia, Korea, the Middle East and New Zealand [60]. One of them was developed by Svante [61]. Started in 2007, Svante designed and built a CO_2 capture facility to capture half a tonne of carbon per day. This technology captures carbon dioxide from flue gas, concentrates it, then releases it for safe storage or industrial use, in 60 s. Today, Svante has several industrial-scale carbon capture projects and collaborations.

Another application example is Air Products. This company possesses solutions for CO_2 capture from fossil fuel conversion before it reaches the atmosphere. The technology has designed and constructed a large-scale system to capture CO_2 from steam methane reformers, which are located within the Valero Refinery in Port Arthur (TX, USA). Air Products has a technology that already separate, purify and transport CO_2 from natural gas reforming, management of syngas from gasification, and oxyfuel combustion in markets such as steel and glass [62].

The Global CCS Institute provides a database of CCUS facilities operating in the world. This organization establishes as large-scale integrated CCSU facilities in its database comprising the capture, transport, and storage of CO_2 at a scale of at least 800 kt of CO_2 annually for a coal-based power plant, or at least 400 kt of CO_2 annually for other emission-intensive industrial facilities (natural gas-based power generation is included). The remaining facilities and initiatives in the database are mentioned as in advancement/deployment status [63]. The last update of the database refers to October 2019.

Currently, there are several CCS facilities in Europe, and they can be classified in three different classes:

(1) Commercial Carbon Capture and Storage Facilities–CO_2 can be captured and transported to be permanently stored; have economic lives similar to the host facility whose CO_2 is captured; must support a commercial return while operating and meet regulatory requirements;
(2) Carbon Capture and Storage Hubs–Commercial facilities although not having a full-chain (capture, transport and storage) operation; several models are considered, combining multiple capture facilities, or CO_2 transport and storage;
(3) Pilot and Demonstration Facilities–CO_2 is captured for testing, developing or demonstrating CCS technologies/processes; CO_2 captured may or may not be transported for permanent storage; A commercial return during operation is not expected.

Taking into account this classification, actually in Europe (accessed data at 1st of April 2021), there are 13 commercial CCS Facilities, two CCS Hubs (one in The Netherlands and another one in United Kingdom), and 29 Pilot and Demonstration facilities [63]. Figure 8 shows the map of the worldwide distribution of CCUS facilities, focusing on Europe.

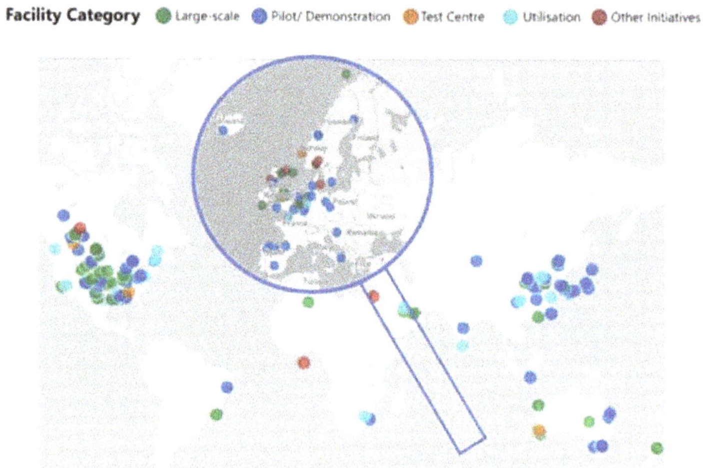

Figure 8. Worldwide distribution of CCUS facilities divided by categories, expanded in Europe (adapted from Institute [63]).

Table 2 summarises the commercial CCS facilities that are working in Europe. Other facilities are under study in test centers or the pilot or demonstration phase. Table A1 to Table A4 (in the Appendix A) summarise these facilities.

Table 2. Summary of large-scale commercial CCS facilities that are working in Europe (Notes: Status: ED—Early Development; AD—Advanced Development; O—Operational; C—Completed; In C—In Construction; Data: represents the starting year of the project; Mtpa—Million tonnes per annum; tpa—tonnens per annum; tpd—tonnes per day).

Name	Status	Country	Data	Industry	Observations
Acorn Scalable CCS Development	ED	UK	2020s	Oil Refining	Scale-up of the pilot project Acorn (Minimum Viable CCS Development)
Caledonia Clean Energy	ED	UK	2024	Power generation	CO_2 capture would be 3 Mtpa and transported via re-purposed pipeline for geological storage in the North Sea of Scotland
Drax BECCS Project	ED	UK	2027	Power generation	Aims to capture 4 Mtpa from one (660 MW) of the biomass-fired power lines at the UK's biggest power station by 2027
Ervia Cork CCS	ED	Ireland	2028	Power generation and refining	CO_2 captured initially from the two-modern gas-fired, combined-cycle gas turbine power stations and Ireland's only oil refining business; transported via a pipeline network to sites in the Kinsale Gas Field
Hydrogen to Humber Saltend (H2H)	ED	UK	2026–2027	Hydrogen production	H2H Saltend is in development to produce blue hydrogen via a new build 600 MW autothermal reformer to decarbonize Triton Power's gas-fired power plant; up to 1.4 million tonnes of CO_2 will be captured.
Hydrogen 2 Magnum (H2M)	ED	Netherlands	2004	Power generation	H2M produce hydrogen to be used in gas power plant in Eemshaven, Germany, Equinor, Vattenfall and Gassunie

Table 2. *Cont.*

Name	Status	Country	Data	Industry	Observations
HyNet North West	ED	UK	Mid 2020s	Hydrogen Production	CO_2 is planned to be captured from the Hydrogen Production & Carbon Capture plant, and transported, together with captured CO_2 from existing nearby industrial sites
Langskip CCS–Fortum Oslo Varme	AD	Norway	2024	Waste Incineratiom	It is in construction to capture about 0.4 Mtpa of CO_2 by 2024 from its cement production plat in southern Norway; the offshore Aurora area has been evaluated as original storage site and will involve a combined ship and pipeline transportation system.
Net Zero Teesside	ED	UK	2020s	Various	Cluster of leading energy-intensive companies to examine the opportunity to build one of Europe's first CCS equipped industrial zones in Tees Valley, UK; starts with a capture capacity of 0.8 Mtpa, that could grow up to 10 Mtpa; CO_2 transported via pipeline to an offshore site in the North Sea
Northern Gas Network H21 North of England	AD	UK	2026	Hydrogen production	H21 aims to convert the UK gas grid from natural gas (methane) to zero-carbon hydrogen
Norway Full Chain CCS	AD	Norway	2023–2024	Various	Aim of 0.8 Mtpa; Capture CO_2 studies are being undertaken by two proponents involved in cement production and a waste-to-energy recovery plant, both in southern Norway; CO_2 would be transported via ship and pipeline to an offshore in the Smeaheia area
Port of Rotterdam CCUS Backbone Initiative (Porthos)	AD	Netherlands	2023	Various	The ambition is to store 2 Mtpa from 2023 on, a total that will run up to 5 Mtpa by 2030
Sleipner CO_2 Storage	O	Norway	1996	Natural Gas Processing	The Spleipner CO_2 storage facility was the first (since 1996) in the world to inject CO_2 into a dedicated geological storage (located offshore in Norway); Approximately 0.85 Mtpa is injected and over 17 Mtpa has been injected since inception to 2019
Snohvit CO_2 Storage	O	Norway	2008	Natural Gas Processing	CO_2 is captured at an LNG facility on the island of Melkoya, Norway; designed to capture 0.7 Mtpa, and CO_2 is transported via pipeline back to the Snohvit field offshore, where more than 4 Mtpa has been stored to date since 2008
The Clean Gas Project	ED	UK	2025	Power generation	Natural gas will be used to generate power via a Combined Cycle Gas turbine gas-fired generation station, with CO_2 captured and transported by pipeline for storage in a formation under the Southern North Sea

As demonstrated with the set of works in development presented in Table 2, great efforts have been made to apply capture, storage, and utilization processes in plants. However, there are only a few large-scale CCS plants in operation in Europe so far. This is related to a series of obstacles preventing this technology from being adopted more widely. In most European countries, the nature of the challenges can be political, economic, technical, and social [64–67].

- Political: lack of political commitment with CCS by some member states;

- Economic: high investment, high operational costs, lack of competitiveness compared with other low carbon technologies; no financial compensation for the additional capital and operating costs associated with CCS; Long-term funding commitments from various public and private sources ensure the continuity of research programs which are necessary for the development of CO_2 utilization;
- Technical: lack of infrastructures for transport and storage;
- Social: CCS is unknown for the overall public; resistance to CO_2 storage concept; environmental risks concerning health and water pollution.

Efforts have been made to combat these barriers. As mentioned, to reach the targets defined in the Paris Agreement, immediate/prompt action would be required to reduce CO_2 emissions. Processes related to CCS can be classified as carbon-positive, near carbon-neutral, or carbon-negative. Carbon positive corresponds to the majority of the processes, which still emit CO_2 for the atmosphere. Carbon-neutral and carbon-negative emissions are responsible for zero carbon emissions (neutral) and CO_2 emissions reduction to the atmosphere. Examples of the "negative" processes able to capture CO_2 are Bioenergy with Carbon Capture and Storage (BECCS) and Direct Air Capture (DAC).

4. Climate Positive Solutions

4.1. Bioenergy with Carbon Capture and Storage (BECCS)

Bioenergy has always been present in the world and used by humans to produce heat. Bioenergy is used in vehicles as fuel (bioethanol) and provides electricity by burning biomass [14]. BECCS is part of the broader CCS technology and is emerging as one of the most advanced technologies to decarbonize emission-intensive industries and sectors and enable negative emissions [14]. This is a group of different technologies to produce energy from biomass and CO_2 storage.

This process using biomass as a fuel source because biomass feedstock draws down CO_2 from the atmosphere through photosynthesis. Biomass is burned (combusted or converted) to biofuel, using digestion or fermentation processes. The heat generated can be used for electricity generation or industrial applications, such as cement, pulp and papermaking, waste incineration, steel and iron, and petrochemical. Conversion leads to gaseous (when digestion is applied) or liquid (when fermentation occurs) fuels production. In the liquid case, it leads to the production of bioethanol. Then, CO_2 is captured from a biomass energy conversion and permanently stored in a suitable geological formation. At the end of the process, the CO_2 emitted during bioenergy production, the CO_2 transported, converted, and utilized should be lower than the CO_2 stored to achieve the primary target-negative emissions. Figure 9 presents a scheme of the BECCS.

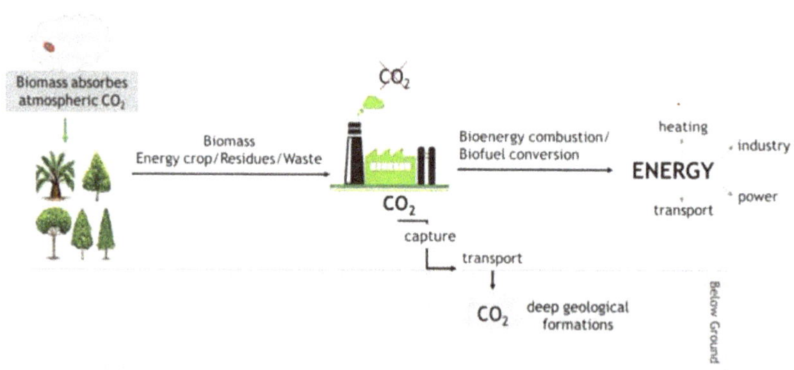

Figure 9. Scheme of bioenergy and carbon capture and storage (BECCS) (adapted from Consoli [14]).

The Global CCS Institute (2019 data) [14] reports the existence of five facilities (one large-scale and four small-scale) actively operating using BECCS technologies worldwide. Approximately 1.5 million tonnes of CO_2 per annum (Mtpa) are captured by these facilities. Table 3 shows a description of these BECCS facilities and the planned projects.

Table 3. Brief description of BECCS facilities operating today and planned projects (Notes: Mtpa—million tonnes per annum; tpa—tonnes per annum; tpd—tonnes per day).

Operating Today—Five Facilities in USA
Illinois CCS (USA)—**1 Mtpa**
Ethanol is produced from corn at its Decatur plant, producing CO_2 as part of the fermentation process
Kansas Arkalon (USA)—**200,000 tpa**
CO_2 is compressed and piped from an ethanol plant in Kansas to Booker and Farnsworth Oil Units in Texas for EOR
Bonanza CCS (USA)—**100,000 tpa**
CO_2 is compressed and piped from an ethanol plant in Kansas to nearby Stewart Oil field for EOR
Husky Energy CO_2 Injection (Canada)—**250 tpd**
CO_2 is compressed and trucked from an ethanol plant (Saskatchewan) to nearby Lashburn and Tangleflags oil fields for EOR
Farnsworth (USA)—**600,000 tonnes**
CO_2 is compressed from an ethanol plant (Kansas) and fertiliser plant (Texas) and piped to Farnsworth oil field for EOR
Planning—One facility in Asia and Two facilities in Europe
Mikawa Power Plant (Japan)
Retrofit of a 49-MW unit power plant (Omuta, Fukuoka Prefecture) to accept 100% of tonne of biomass with a CO_2 capture facility.Current situation: identify a secure offshore storage site
Drax Power Plant (UK)
Biomass power generation pilot (North Yorkshire): high potential to develop CO_2 capture and storage
Drax Power Plant (UK)
BECCS integration into waste-to-energy and a cement plants: Plant: plans to capture 400,000 tpa of CO_2 (Klemetsrud waste-to-energy)Currently co-fires up to 30% biomass and plans to capture up to 400,000 tpa of CO_2 (Norcem Cement plant) CO_2 will be sent to a storage site (Norwegian North Sea) from waste-to-energy and cement plants

Note: Mtpa is million tonnes per annum of CO_2; tpa is tonnes per annum of CO_2; tpd is tonnes per day of CO_2.

4.2. Direct Air Capture (DAC)

Industrial applications containing air capture technology are not new and have existed since 1930 [68].

In contrast to carbon dioxide capture from sources, such as cement or biomass plants, direct air capture (DAC) is a technology that captures CO_2 directly from the ambient air and generates an enriched stream of CO_2 for storage or use. The process can be denominated as physical or chemical separation. Figure 10 presents a scheme of the DAC process.

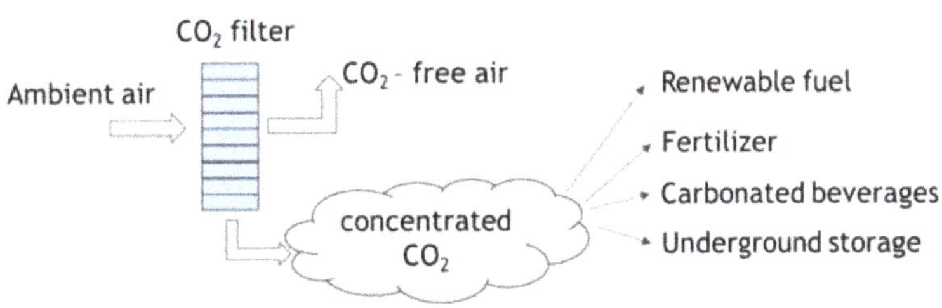

Figure 10. Scheme of DAC (adapted from Gutknecht [69]).

It is common to divide the DAC process into three different classes, regarding the approach to separate CO_2 from the air: chemical, cryogenic, and membranes [70].

Two technology approaches are being used to extract CO_2 from the atmosphere in the chemical systems: liquid systems (liquid solvents) and solid systems (solid sorbents) direct air capture. In the cryogenic processes, CO_2 is removed from the air by freezing as a by-product of cryogenic oxygen separation. Membranes are used to separate CO_2 from the air and seawater. Chemical systems are the preferred processes of DAC used by companies.

Liquid systems pass air through chemical solutions, which removes the CO_2 while returning the rest of the air to the environment. For example, a typical process used is when sodium hydroxide is the solvent applied (used in the pulp and paper industry). CO_2 reacts with sodium hydroxide (NaOH) and precipitates sodium carbonate (Na_2CO_3), which produces a highly pure gaseous CO_2 stream when heated; after that, sodium hydroxide is recycled from sodium carbonate.

The reaction occurs between NaOH and CO_2, as presented in Equation (2):

$$2\text{NaOH}(\text{solution}) + \text{CO}_2 \rightarrow \text{Na}_2\text{CO}_3(\text{solution}) + \text{H}_2\text{O} \tag{2}$$

$$\Delta\text{H} = -105\ \text{kJ/mol}$$

This process has a high potential to obtain high loadings of CO_2 over a wide range of operating conditions and system designs because of the strong binding energy associated with the reaction presented in Equation (2). A disadvantage is the high energy requirements for releasing the CO_2 during the regeneration stage [20].

Figure 11 shows a brief scheme of this process.

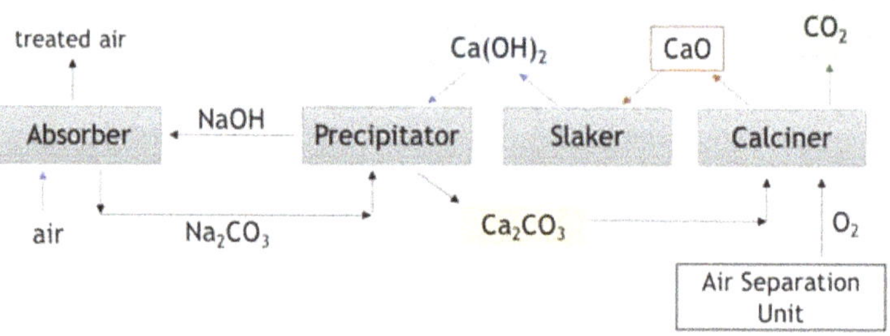

Figure 11. Brief scheme of a liquid solvent process used for capturing CO_2 from air, using NaOH as the absorber (adapted from Mazzotti et al. [71]).

Solid direct air capture technology makes use of solid sorbent filters that chemically bind with CO_2. When the filters are heated, they release the concentrated CO_2, which can be captured for sequestration or utilization.

However, CO_2 in the air is approximately 300 times (~400 ppm) more dilute than in flue gas from a coal-fired power plant, which results in a costly process to separate CO_2 with the same end purity as the one obtained in the CO_2 captured from fossil fuel power plants [72]. Figure 12 shows a brief scheme of the DAC process.

At present, few companies are involved in the DAC field, all designing or using different technologies of DAC, and different markets are focused.

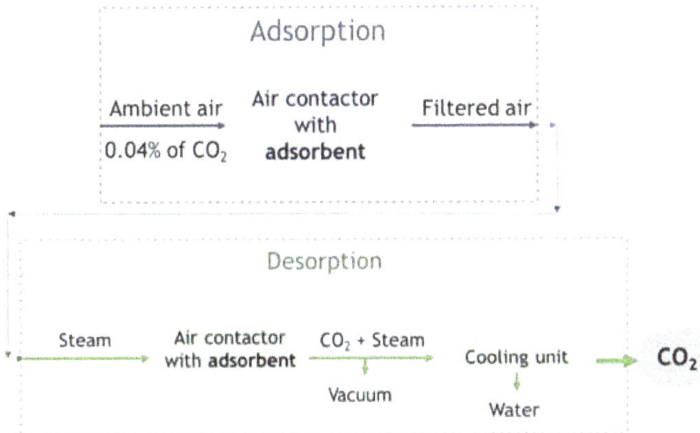

Figure 12. Schematic representation of the DAC process (adapted from Sinha and Realff [73]).

Carbon Engineering Ltd. (CE, Vancouver, BC, Canada) uses liquid alkali metal oxide sorbents regenerated by heat at around 800 °C. CE uses natural gas to power its machines, co-capturing CO_2 from the flue gas stream of the burned natural gas in addition to atmospheric capture [74,75].

Global Thermostat (GT, New York, NY, USA) is a US company which uses a solid amine-based sorbent material for CO_2 capture from air, regenerated at around 80–100 °C [76].

Also, using DAC design, Climeworks AG (Zurich, Switzerland) capture CO_2 with a system based on an adsorption-desorption process with alkaline-functionalized adsorbents. The adsorption is performed at ambient conditions while the desorption occurs using a temperature-vacuum-swing (TVSA) process. The pressure decrease and the temperature increase from 80 to 120 °C, allow to release the CO_2 [76]. The enriched stream of CO_2 is produced at 1 bar with a purity of >99.8%. If the relative humidity on the feed is high, the H_2O is also extracted from the air as a by-product [77]. The first commercial DAC plant was presented in 2017 in Switzerland from Climeworks, with a capacity for 900 t of CO_2 captured per year from the air.

Currently, in Europe, in the United States (US), and in Canada, there are more than 15 DAC plants operating worldwide, most of them are small and sell CO_2 captured for use (in carbonated drinks, for example). However, the first large-scale DAC plant has been developed in the US by a Carbon Engineering Ltd. and Occidental Petroleum partnership. The plant will capture up to 1 $MtCO_2$ (metric tonnes of CO_2) per year for EOR. This unit could become operational as early as 2023 [76]. Table 4 presents the companies that are working to commercialize DAC systems nowadays.

Several studies have been presented with direct air capture applications to obtain climate change mitigation, some more optimistic than others. Creutzig, et al. [78] estimates that DAC will reach 1 Gt of CO_2 per year in 2050. Fasihi, et al. [79] presents an estimative of about 7 Gt of CO_2 captured per year in the energy system, and about 8 Gt of CO_2 captured per year in carbon dioxide removal in the same year.

Today, the costs involved in direct air capture systems are approximately 510 € per tonnes of CO_2 captured [80].

The transition to a net-zero energy system, in which the amount of CO_2 released to the atmosphere is equivalent to the amount being removed, is highly dependent on the carbon removal processes. The application of decarbonization strategies in the several sectors as aviation and heavy industry would be very difficult. In these cases, carbon removal technologies can be the key for an effective transition. In the 2030 Sustainable Development Scenario, it was defined that CO_2 capture by direct air capture should reach almost 10 Mt of CO_2 per year (in 2030) [76].

Table 4. Companies Working to Commercialize Systems of Direct Air Capture technology [72].

Company	Type of System	Type of Technology	Type of Regeneration	Purity/ Application	Scale
Carbon Engineering Ltd.	Liquid solvent	Potassium hydroxide solution/calcium carbonation	Temperature	99%	Pilot 1 tonne per day
Climeworks	Solid sorbent	Amine-functionalized filter	Temperature or vacuum	99%w/dilution depending on the application	Demonstration 900 tonne per year
Global Thermostat	Solid sorbent	Amine-modified monolith	Temperature and/or vacuum	99%	1000 tonne per year
Infinitree	Solid sorbent	Ion-exchange sorbent	Humidity	3–5% algae	Laboratory
Skytree	Solid sorbent	Porous plastic beads functionalized with benzylamines	Temperature	Air purification, greenhouses	Appliance

5. Industrial Processes

Industrial processes are responsible for raw materials conversion into useable products, which results in energy consumption and CO_2 emissions. For this conversion, fossil fuels continue to satisfy most of the industrial energy demand. However, these processes can be transformed to meet global climate changes. The industrial CO_2 emissions can be categorized into four main groups [27]:

- Energy-related emissions: combustion of coal, oil, and natural gas (considering biomass with an emission factor of zero);
- Process emissions: associated with chemical and physical reactions, such as the production of aluminum, ferroalloys, lubricants and paraffins, and fuels through coal and gas-to-liquid processes, etc.;
- Direct emissions: all emissions associated with industrial processes, except the electricity, heat, and steam purchases (energy-related emissions plus process emissions);
- Indirect emissions: all emissions "out of the facilities", including electricity, heat and steam purchased.

Industry is responsible for about one-quarter of CO_2 emissions from energy and industrial processes, being 90% of the direct GHG emissions from industrial production, and 40% of global energy demand, especially in cement, in iron and steel and in chemicals industries, which are the most challenging for emissions reduction. Between 1990 and 2017, industrial CO_2 emissions increased 70%. According to Clean Technology Scenario (CTS), consistent with the Paris Agreement defined targets, more than 28 Gt of CO_2 must be captured from industrial facilities until 2060 [27]. CCUS can be a critical factor in the industry decarbonization action. CCUS technologies will contribute to a reduction of 21 Gt of CO_2 of emissions (27%) in the period of 2017–2060, from the cement (18%, capturing 5 Gt of CO_2), iron and steel (15%, capturing 10 Gt of CO_2), and chemical subsectors (38%, capturing 14 Gt of CO_2) [27].

Several industrial sectors produce CO_2 at different temperatures, concentrations, purities, pressures, and volumes, and for all of them, carbon dioxide capture technologies could be applicable. This will be vital for energy-intensive industries, such as those listed below to capture carbon if the EU is to reach its climate targets. These industry subsectors consider iron and steel, chemicals and petrochemicals, cement, pulp and paper, aluminium, and other industries such as ceramics and glass production.

The costs involved in CO_2 capture vary greatly by point source and by capture technology. Costs range from 15 USD per tonne of carbon dioxide (USD/t of CO_2) to 60 USD/t of CO_2 for concentrated CO_2 streams (e.g., natural gas processing and bioethanol production through fermentation), or from 40 USD/tCO_2 to 80 USD/tCO_2 for coal- and gas-fired power plants. The costs can be over USD 100/t of CO_2 for smaller or more dilute point sources (e.g., industrial furnaces) [27].

The current status of industrial sectors is encouraging. Several works are in development today, especially involving the CO_2 capture from high -purity CO_2 sources.

6. Conclusions

International climate obligations, especially the values established in the Paris Agreement, require detailed monitoring and reporting of greenhouse gas emissions, which allowed to observe the increase of CO_2 emissions over time. These scary numbers allow understanding the immediate need to act to reduce the emissions. In this regard, CO_2 capture, utilization, and storage have demonstrated a high potential to be used to reduce global warming potential from power plants. CCS/CCUS has many challenges to overcome. For that, continuous advancement of knowledge is essential to improve the economic and environmental feasibility and technologies potential. As can be seen, several projects are under study to improve the capture of CO_2 and utilization/storage. In the future, some technologies may offer a range of potential opportunities for a sustainable global industry, supporting the climate change objectives, the circular economy, renewable energy deployment, the evolution of CO_2 capture systems, among others. With the 1.5 °C pathway of the Paris Agreement, CO_2 emissions should be 0 in 2050. Therefore, there is still a long path ahead.

Author Contributions: Conceptualization, M.J.R.; Writing—original draft preparation, M.J.R.; Writing—review and editing, M.J.R., A.P., A.F.P.F., A.M.R. and A.E.R.; Visualization, M.J.R.; Supervision, A.F.P.F., A.M.R. and A.E.R. All authors have read and agreed to the published version of the manuscript.

Funding: This work was financially supported by: Base Funding-UIDB/50020/2020 of the Associate Laboratory LSRE-LCM-funded by national funds through FCT/MCTES (PIDDAC). Financial support o NORTE-01-0145-FEDER-000006f FCT–Fundação para a Ciência e Tecnologia under CEEC Institucional program is also acknowledged.

Institutional Review Board Statement: Not applicable.

Informed Consent Statement: Informed consent was obtained from all subjects involved in the study.

Conflicts of Interest: No conflict of interest.

Nomenclature

AMP	2-Amino-2-methyl-1-propanol
BECCS	Bioenergy with Carbon Capture and Storage
CCS	Carbon Capture and Storage
CCU	Carbon Capture and Utilization
CCUS	Carbon Capture, Utilization, and Storage
DAC	Direct Air Capture
DEA	Diethanolamine
EGR	Enhanced Gas Recovery
EOR	Enhanced Oil Recovery
ESA	Electrical Swing Adsorption
HBCC	Hydrate-based Carbon Dioxide Capture
IGCC	Integrated Gasification Combined Cycle
MDEA	Methyl diethanolamine
MEA	Monoethanolamine
PSA	Pressure Swing Adsorption
TSA	Temperature Swing Adsorption
TVSA	Temperature-Vacuum Swing Adsorption

Appendix A

Table A1. Summary of other CCS facilities that are working in Europe.

Name	Status	Country	Data	Industry	Observations
CATO Programme	O	The Netherlands	2004	Various	Responsible for covering the full CCS chain and addressing both fundamental and applied topics including regulation and safety and public perception
CO_2 FieldLab Project	C	Norway	-	N/A	Project is led by Sintef Petroleum Research, with the purpose of testing the sensitivity of a variety of monitoring systems by observing the migration of small amounts of injected CO_2 in the shallow subsurface
CO_2 MultiStore Joint Industrial Project (JIP)	C	UK	2012	N/A	Project is led by Scottish Carbon Capture and Storage with joint funding and expert support from CCS project developers and public corporations, to support the development of multi-use regional CO_2 storage assets
Hisarma Pilot Plant (Reducing CO_2 Emissions in Steelmaking)	O	The Netherlands	2007	Iron and Steel Production	A coal-based Hisarna smelting reduction process in steelmaking industry was developed by Tata Steel, Rio Tinto and ULCOS partners, and it has been operational since 2011. CO_2 emissions were reduced by 20%, and can be 80% lower when CCS is applied
QICS Project	C	UK	2010	N/A	Project led by the Plymouth Marine Laboratory, to quantify and monitor Environmental Impacts of Geological Carbon Storage involved in the assessment and monitoring of the first controlled release of CO_2 into seaed sediments

Table A2. Summary of pilot and demonstration mode CCS facilities in Europe.

Name	Status	Country	Data	Industry	Observations
ELCOGAS Pre-combustion Carbon Capture Pilot Project: Puertollano	C	Spain	2010	Power Generation	A pilot plant was integrated into the Puertollano IGCC plant in Spain to test the feasibility of pre-combustion technology to capture CO_2 in an IGCC environment that uses solid fossil fuels and wastes as feedstock; operational tests occurred in 2010/2011
Aberthaw Pilot Carbon Capture Facility	C	UK	2013	Power generation	A pilot-scale plant at the Aberthaw power station in South Wales UK tested the Cansolv integrated CO_2 and SO_2 removal system during 2013/2014
Abu Dhabi CCS (Phase 1 being Enirates Steel Industries)	O	United Arab Emirates	2016	Iron and Steel Production	First fully commercial CCS facility in the iron and steel industry, and involves the CO_2 capture via a new build CO_2 Compression facility using high purity CO_2 produced as a by-product of the direct reduced iron-making process at the Emirates Steel Industries factory in Mussafah. The compression facility has a capture capacity of 0.8 Mtpa. The CO_2 is captured and is transported via pipeline to Abu Dhabi National Oil Company ADNOC oil reservoirs for EOR
Acorn (Minimum Viable CCS Development)	AD	United Kingdom	2021–2022	Various	Initiate a low cost full chain CCS project in the North East of Scotland; cluster of capture, transport and storage infrastructure; CO_2 is separated from natural gas and vented, adjacent to an offshore transport pipeline, which connects to a well understood offshore basin, rich in storage opportunities

Table A2. *Cont.*

Name	Status	Country	Data	Industry	Observations
Brindisi CO_2 Capture Pilot Plant	C	Italy	2010	Power Generation	A pilot-scale plant at the Brindisi power plant in south-eastern Italy tested a number of solvent technologies during 2010–2012
Buggenum Carbon Capture (CO_2 Catch-up) Pilot Project	C	The Netherlands	2011	Power Generation	A pilot-scale plant at the Willem-Alexander power plant in The Netherlands (now closed) undertook a CO_2 capture testing and R&D program between 2011 and 2013
C2A2 Field Pilot-Le Havre	C	France	2013	Power Generation	A pilot-scale plant at the Le Havre power plant in France tested a specific carbon capture technology during parts of 2013 and 2014
CarbFix Project	O	Iceland	2012	Power Generation	Study of injection of pure CO_2 and a gas mixture of CO_2 and H2S, dissolved in water, into basaltic formations; pilot tests in 2012 injected over 200 tonnes of CO_2 from a geothermal power plant
CASTOR	C	Denmark	2006	Power Generation	Tests of three solvents in a post-combustion pilot plant located at the coal-fired Esbjerg power plant in Denmark
CEMCAP	C	Norway	2015	Cement Production	Prepare the ground for large-scale implementation of CO_2 capture in the European cement industry; designed to strengthen and complement the Norcem and ECRA CCS projects; technical development, including technology demonstration in a simulated industrial environment.
CESAR	C	Denmark	2008	Power Generation	Project for the post-combustion capture work undertaken at the coal-fired Esbjerg pilot plant in Denmark under the CASTOR project; modifications to the Esbjerg pilot plant were undertaken during 2008, after which a three test campaign was conducted covering a benchmark and two novel solvents
CIUDEN: CO_2 Capture & Transport Technology Development Plant	C	Spain	2012	Power Generation	The Hontomin Technology Development Plant—CO_2 Capture & Transport the CIUDEN Technology Development Center, successfully completed the full CO_2 capture process, using oxy-combustion in the circulating fluidized bed CFB boiler with the compression and purification unit
CIUDEN: CO_2 Storage Technology Development Plant	O	Spain	2015	N/A	The Hontomin Storage Technology Development Plant—the site includes one injection well and a monitoring well;10,000 tonnes of CO_2 are planned to be injected in the period 2017–2020
CO_2 Capture Test Facility at Norcem Brevik	C	Norway	2013	Cement Production	A real cement flue gas at the Brevik plant in Norway was used to test three different post-combustion technologies, while investigations on a fourth technology were performed offsite based on a pilot installed at Stuttgart University; tests were done from 2013 to 2017 and aimed to demonstrate the CO_2 capture from a cement plant, to improve understanding of these technologies for large-scale application
DMX™ Demonstration in Dunkirk	AD	France	2022	Iron and Steel Production	Designed by Axens, started in 2020 at the ArcelorMittal steelworks site in Dunkirk; able to capture 0.5 metric tonnes of CO_2 an hour from steelmaking gases by 2022

Table A2. *Cont.*

Name	Status	Country	Data	Industry	Observations
Drax bioenergy carbon capture pilot plant	O	United Kingdom	2019	Power Generation	The CO_2 capture pilot plant captures 1 tpd from the Drax power station unit, which runs 100% biomass feedstock
ELCOGAS Pre-combustion Carbon Capture Pilot Project: Puertollano	C	Spain	2010	Power Generation	A pilot plant was integrated into the Puertollano IGCC plant in Spain to test the feasibility of pre-combustion technology to capture CO_2 in an IGCC environment that uses solid fossil fuels and wastes as the main feedstock
Ferrybridge Carbon Capture Pilot (CCPilot100+)	C	United Kingdom	2011	Power Generation	Involves the capture of 100 tpd of CO_2 from a flue gas stream at the Ferrybridge power station; designed to test the application of an amine-based, post-combustion capture process under realistic operating conditions
Geothermal Plant with CO_2 Re-injection	C	Croatia	2018	Power Generation	A hybrid geothermal system is used, utilizing the energy potential of hot brines with dissolved natural gases to deliver combined heat and power at its Draškovec development; is expected to supply 17–18 MW of power; CO_2 separation, capture and injection capacity is at around 50,000 tpa
K12-B CO_2 Injection Project	C	The Netherlands	2004	Natural Gas Processing	CO_2 is captured at the offshore natural gas production facility at the K12-B gas field and injected back into the depleted gas reservoir; cumulative injection in 2017 was over 100,000 tonnes
Karlshamn Field Pilot	C	Sweden	2009	Power Generation	One of a number of test facilities used by Alstom to test the viability of its Chilled Ammonia Process for CO_2 capture and involved the capture of around 30 tpd of CO_2
Ketzin Pilot Project	C	Germany	2004	Power Generation and Hydrogen Production	The first geological CO_2 storage project on the European mainland and one of the largest storage pilot projects in the world; constituted by three phases, over 67,000 tonnes of CO_2 were injected between 2008 and 2013, and post-injection and site behavior monitoring completed in 2017
La Pereda Calcium Looping Pilot Plant	C	Spain	2012	Power Generation	In operation in 2012, undertook three European funded 'projects' or test campaigns, testing the viability of post-combustion capture by calcium looping, 1.7 MWth
Lacq CCS Pilot Project	C	France	2010	Power Generation	A storage-focused project of global significance that injected 51,000 tonnes of CO_2 over a 39 month period from 2010 to 2013; including a comprehensive monitoring plan
LEILAC—Low Emissions Intensity Lime and Cement Project	In C	Belgium	2020's	Cement Production	Designed, built and operated a pilot plant to: Direct Separation calcining technology can work at the temperatures necessary to process limestone for the lime and cement industries; capture over 95% of the CO_2 emissions from both industries without significant energy or capital penalty
Renfrew Oxy-fuel (Oxycoal 2) Project	C	United Kingdom	2007	Power Generation	Consisted of an initial oxy-fuel technology mapping phase followed by testing of a 40-MWth oxy-fuel burner at Renfrew, Scotland, under realistic operating conditions and included testing adaptation from air-firing to oxy-fuel firing on pulverized coal; the pilot facility completed 20 test days

Table A2. *Cont.*

Name	Status	Country	Data	Industry	Observations
Schwarze Pumpe Oxy-fuel Pilot Plant	C	Germany	2008	Power Generation	The 30 MWth Schwarze Pumpe oxy-fuel pilot was the world's first large-scale testing of the entire oxy-fuel combustion technology chain; In 2014, it was discontinued research into coal-fired power with CCS; the pilot plant captured and liquefied 11,000 tonnes of CO_2; Around 1500 tonnes of CO_2 from the Schwarze Pumpe oxy-fuel capture pilot plant were injected
STEPWISE Pilot of SEWGS Technology at Swerea/Mefos	O	Sweden	2017	Iron and Steel Production	The Sorption Enhanced Water Gas Shift reaction SEWGS process is to be demonstrated at a CO_2 capture rate of 14 tonnes per day; the pilot plant is fed with blast furnace gas from the adjacent steel plant of SSAB; the test facility was launched in September 2017
Wilhelmshaven CO_2 Capture Pilot Plant	C	Germany	2012	Power Generation	CO_2 capture from a side stream of the Wilhelmshaven coal-fired power station; designed to capture 70 tpd of CO_2 at full capacity; achieved 4500 h of operation in the first quarter of 2014

Table A3. Summary of CCS facility tests centers in Europe.

Name	Status	Country	Data	Industry	Observations
Technology Centre Mongstad (TCM)	O	Norway	2012	Oil Refining	The demonstration test facility comprises two capture units, one designed for amine-based solvents and the other for chilled aqueous ammonia
UKCCSRC Pilot-scale Advanced Capture Technology (PACT)	O	United Kingdom	-	Power Generation	PACT facilities bring together a range of integrated pilot-scale and accompanying specialist research and analytical facilities, supported by leading academic expertise; post-combustion pilot is installed and is jointly operated by the Universities of Leeds and Sheffield

Table A4. Summary of CO_2 utilization facilities in Europe.

Name	Status	Country	Data	Industry	Observations
ArcelorMittal Steelanol Ghent	In C	Belgium	2020	Iron and Steel Production	ArcelorMittal Steelanol Ghent
Port Jérôme CO_2 Capture Plant	O	France	2015	Hydrogen Production	Port Jérôme CO_2 Capture Plant
Twence Waste-to-energy CO_2 Capture and Utilisation	O	The Netherlands	2014	Waste Incineration	Twence Waste-to-energy CO_2 Capture and Utilisation

References

1. Climate Change Service. Surface Air Temperature for September 2019. Available online: https://climate.copernicus.eu/surface-air-temperature-september-2019 (accessed on 14 March 2021).
2. Borowski, P.F. Nexus between water, energy, food and climate change as challenges facing the modern global, European and Polish economy. *AIMS Geosci.* **2020**, *6*, 397–421. [CrossRef]
3. Folger, P. *Carbon Capture: A Technology Assesssment*; Congressional Research Service: Washington, DC, USA, 2013; p. 3.
4. Lallanilla, M. Greenhouse Gas Emissions: Causes & Sources. Available online: http://www.livescience.com/37821-greenhouse-gases.html (accessed on 26 May 2016).
5. Jefferson, M. Energy policies for sustainable development. In *World Energy Assessment: Energy and the Challenge of Sustainability*; Communications Development Incorporated: Washington, DC, USA, 2000.
6. Green, C.; Byrne, K. Biomass: Impact on Carbon Cycle and Greenhouse Gas Emissions. *Encycl. Energy* **2004**, *1*, 223–236. [CrossRef]
7. Quadrelli, R.; Peterson, S. The energy-climate challenge: Recent trends in CO_2 emissions from fuel combustion. *Energy Policy* **2007**, *35*, 5938–5952. [CrossRef]

8. Olajire, A.A. CO_2 capture and separation technologies for end-of-pipe applications—A review. *Energy* **2010**, *35*, 2610–2628. [CrossRef]
9. Masson-Delmotte, V. *Global Warming of 1.5 °C.*; IPCC—Intergovernmental Panel on Climate Change: Geneva, Switzerland, 2019; ISBN 978-92-9169-153-1.
10. Masson-Delmotte, V.P.; Zhai, H.-O.; Pörtner, D.; Roberts, J.; Skea, P.R.; Shukla, A.; Pirani, W.; Moufouma-Okia, C.; Péan, R.; Pidcock, S.; et al. *Global Warming of 1.5 °C. An IPCC Special Report on the Impacts of Global Warming of 1.5 °C above Pre-Industrial Levels and Related Global Greenhouse Gas Emission Pathways, in the Context of Strengthening the Global Response to the Threat of Climate Change, Sustainable Development, and Efforts to Eradicate Poverty*; IPCC: Geneva, Switzerland, 2018.
11. Foley, A.; Smyth, B.M.; Pukšec, T.; Markovska, N.; Duić, N. A review of developments in technologies and research that have had a direct measurable impact on sustainability considering the Paris agreement on climate change. *Renew. Sustain. Energy Rev.* **2017**, *68*, 835–839. [CrossRef]
12. Sekera, J.; Lichtenberger, A. Assessing Carbon Capture: Public Policy, Science, and Societal Need. *Biophys. Econ. Sustain.* **2020**, *5*, 14. [CrossRef]
13. Bui, M.; Adjiman, C.S.; Bardow, A.; Anthony, E.J.; Boston, A.; Brown, S.; Fennell, P.S.; Fuss, S.; Galindo, A.; Hackett, L.A.; et al. Carbon capture and storage (CCS): The way forward. *Energy Environ. Sci.* **2018**, *11*, 1062–1176. [CrossRef]
14. Consoli, C. *Bioenergy and Carbon Capture and Storage*; Global CCS Institute: Docklands, Australia, 2019.
15. International Energy Agency. Global Emissions in 2019. Available online: https://www.iea.org/articles/global-co2-emissions-in-2019 (accessed on 18 March 2021).
16. International Energy Agency. *Global Energy Review 2020*; International Energy Agency: Paris, France, 2020. Available online: https://www.iea.org/reports/global-energy-review-2020 (accessed on 18 March 2021).
17. Tiseo, I. Global Distribution of CO_2 Emissions from Fossil Fuel and Cement by Sector 2020. Available online: https://www.statista.com/statistics/1129656/global-share-of-co2-emissions-from-fossil-fuel-and-cement/ (accessed on 18 March 2021).
18. Climate Action Tracker. 2100 Warming Projections. Available online: https://climateactiontracker.org/global/temperatures/ (accessed on 18 April 2021).
19. CarbonBrief—Clear on Climate. Explainer: The high-emissions 'RCP8.5' global warming scenario. Available online: https://www.carbonbrief.org/explainer-the-high-emissions-rcp8-5-global-warming-scenario (accessed on 3 September 2020).
20. Pires, J.C.M.; Martins, F.G.; Alvim-Ferraz, M.C.M.; Simões, M. Recent developments on carbon capture and storage: An overview. *Chem. Eng. Res. Des.* **2011**, *89*, 1446–1460. [CrossRef]
21. Songolzadeh, M.; Ravanchi, M.T.; Soleimani, M. Carbon Dioxide Capture and Storage: A General Review on Adsorbents. *World Acad. Sci. Eng. Technol.* **2012**, *6*, 213–220.
22. International Energy Agency. Carbon Capture, Utilisation and Storage. Available online: https://www.iea.org/fuels-and-technologies/carbon-capture-utilisation-and-storage (accessed on 3 August 2020).
23. Marocco Stuardi, F.; MacPherson, F.; Leclaire, J. Integrated CO_2 capture and utilization: A priority research direction. *Curr. Opin. Green Sustain. Chem.* **2019**, *16*, 71–76. [CrossRef]
24. Romasheva, N.; Ilinova, A. CCS Projects: How Regulatory Framework Influences Their Deployment. *Resources* **2019**, *8*, 181. [CrossRef]
25. Global CCS Institute. *The Global Status of CCS*; Summary report; Global Carbon Capture and Storage Institute Ltd.: Melbourne, Australia, 2015; Available online: https://www.globalccsinstitute.com/wp-content/uploads/2018/12/Global-Status-Report_2015_Summary.pdf (accessed on 3 August 2020).
26. International Energy Agency. *Energy Technology Perspectives*; International Energy Agency: Paris, France, 2017. Available online: https://www.iea.org/topics/energy-technology-perspectives (accessed on 18 March 2021).
27. International Energy Agency. Transforming Industry through CCUS. 2019. Available online: https://www.iea.org/reports/transforming-industry-through-ccus (accessed on 18 March 2021).
28. Dindi, A.; Quang, D.V.; Vega, L.F.; Nashef, E.; Abu-Zahra, M.R.M. Applications of fly ash for CO_2 capture, utilization, and storage. *J. CO2 Util.* **2019**, *29*, 82–102. [CrossRef]
29. IPCC. *IPCC Special Report on Carbon Dioxide Capture and Storage*; Prepared by Working Group III of the Intergovernmental Panel on Climate Change; Metz, B.O., Davidson, H.C., de Coninck, M.L., Meyer, L.A., Eds.; Cambridge University Press: Cambridge, UK; New York, NY, USA, 2005; p. 442.
30. Nota, G.; Nota, F.D.; Peluso, D.; Toro Lazo, A. Energy Efficiency in Industry 4.0: The Case of Batch Production Processes. *Sustainability* **2020**, *12*, 6631. [CrossRef]
31. Global CCS Institute. *CO_2 Capture Technologies*; Global Carbon Capture and Storage Institute: Canberra, Australia, 2012; pp. 1–13.
32. Pastor-Pérez, L.; Baibars, F.; Le Sache, E.; Arellano-García, H.; Gu, S.; Reina, T.R. CO_2 valorisation via Reverse Water-Gas Shift reaction using advanced Cs doped Fe-Cu/Al_2O_3 catalysts. *J. CO_2 Util.* **2017**, *21*, 423–428. [CrossRef]
33. Rackley, S.A. *Carbon Capture from Power Generation*, 2nd ed.; Carbon Capture and Storage; Elsevier: London, UK, 2017; ISBN 9780128120415.
34. U.S. Department Energy. Pre-Combustion CO_2 Capture. Available online: https://www.energy.gov/fe/science-innovation/carbon-capture-and-storage-research/carbon-capture-rd/pre-combustion-carbon (accessed on 15 August 2020).
35. National Energy Technology Laboratory Post-Combustion CO_2 Capture. Available online: https://www.netl.doe.gov/coal/carbon-capture/post-combustion (accessed on 1 April 2021).

36. Buhre, B.J.P.; Elliott, L.K.; Sheng, C.D.; Gupta, R.P.; Wall, T.F. Oxy-fuel combustion technology for coal-fired power generation. *Prog. Energy Combust. Sci.* **2005**, *31*, 283–307. [CrossRef]
37. Wong, S. *Module 3—CO_2 Capture: Pre-Combustion (Decarbonisation) and Oxy-Fuel Technologies*; Global CCS Institute: Docklands, VIC, Australia, 2011; Volume 1, pp. 45–54. Available online: https://www.globalccsinstitute.com/archive/hub/publications/114711/building-capacity-co2-capture-and-storage-apec-region (accessed on 3 August 2020).
38. Armstrong, K.; Styring, P. Assessing the Potential of Utilization and Storage Strategies for Post-Combustion CO_2 Emissions Reduction. *Front. Energy Res.* **2015**, *3*. [CrossRef]
39. Bhown, A.S.; Freeman, B.C. Analysis and Status of Post-Combustion Carbon Dioxide Capture Technologies. *Environ. Sci. Technol.* **2011**, *45*, 8624–8632. [CrossRef]
40. Leung, D.Y.C.; Caramanna, G.; Maroto-Valer, M.M. An overview of current status of carbon dioxide capture and storage technologies. *Renew. Sustain. Energy Rev.* **2014**, *39*, 426–443. [CrossRef]
41. Luis, P. Use of monoethanolamine (MEA) for CO_2 capture in a global scenario: Consequences and alternatives. *Desalination* **2016**, *380*, 93–99. [CrossRef]
42. Walspurger, S.; Van Dijk, H.A.J. *EDGAR CO_2 Purity: Type and Quantities of Impurities Related to CO2 Point Source and Capture Technology: A Literature Study*; The Energy Research Centre of The Netherlands: Petten, The Netherlands, 2012.
43. Samanta, A.; Zhao, A.; Shimizu, G.K.H.; Sarkar, P.; Gupta, R. Post-Combustion CO_2 Capture Using Solid Sorbents: A Review. *Ind. Eng. Chem. Res.* **2012**, *51*, 1438–1463. [CrossRef]
44. Grande, C.A. Advances in Pressure Swing Adsorption for Gas Separation. *ISRN Chem. Eng.* **2012**, *2012*, 13. [CrossRef]
45. Regufe, M.J.; Ribeiro, A.M.; Ferreira, A.F.P.; Rodrigues, A. CO_2 Storage on Zeolites and Other Adsorbents. In *Nanoporous Materials for Gas Storage*; Kaneko, K., Rodríguez-Reinoso, F., Eds.; Springer: Singapore, 2019; pp. 359–381.
46. Yang, H.; Li, J.-R. Metal-Organic Frameworks (MOFs) for CO_2 Capture. In *Porous Materials for Carbon Dioxide Capture*; Lu, A.-H., Dai, S., Eds.; Springer: Berlin/Heidelberg, Germany, 2014; pp. 79–113.
47. Regufe, M.J.; Tamajon, J.; Ribeiro, A.M.; Ferreira, A.; Lee, U.H.; Hwang, Y.K.; Chang, J.-S.; Serre, C.; Loureiro, J.M.; Rodrigues, A.E. Syngas Purification by Porous Amino-Functionalized Titanium Terephthalate MIL-125. *Energy Fuels* **2015**, *29*, 4654–4664. [CrossRef]
48. Lu, C.; Bai, H.; Wu, B.; Su, F.; Hwang, J.F. Comparative study of CO_2 capture by carbon nanotubes, activated carbons, and zeolites. *Energy Fuels* **2008**, *22*, 3050–3056. [CrossRef]
49. Regufe, M.J.; Ferreira, A.F.P.; Loureiro, J.M.; Rodrigues, A.; Ribeiro, A.M. Development of Hybrid Materials with Activated Carbon and Zeolite 13X for CO_2 Capture from Flue Gases by Electric Swing Adsorption. *Ind. Eng. Chem. Res.* **2020**, *59*, 12197–12211. [CrossRef]
50. Regufe, M.J.; Ferreira, A.F.P.; Loureiro, J.M.; Rodrigues, A.; Ribeiro, A.M. Electrical conductive 3D-printed monolith adsorbent for CO_2 capture. *Microporous Mesoporous Mater.* **2019**, *278*, 403–413. [CrossRef]
51. Clodic, D.; Younes, M. A new Method for CO2 Capture: Frosting CO2 at Atmospheric Pressure. In Proceedings of the Greenhouse Gas Control Technologies—6th International Conference, Kyoto, Japan, 1–4 October 2002; Gale, J., Kaya, Y., Eds.; Pergamon: Oxford, UK, 2003; pp. 155–160.
52. Tuinier, M.J.; van Sint Annaland, M.; Kramer, G.J.; Kuipers, J.A.M. Cryogenic CO_2 capture using dynamically operated packed beds. *Chem. Eng. Sci.* **2010**, *65*, 114–119. [CrossRef]
53. Nanda, S.; Reddy, S.N.; Mitra, S.K.; Kozinski, J.A. The progressive routes for carbon capture and sequestration. *Energy Sci. Eng.* **2016**, *4*, 99–122. [CrossRef]
54. Hart, A.; Gnanendran, N. Cryogenic CO_2 capture in natural gas. *Energy Procedia* **2009**, *1*, 697–706. [CrossRef]
55. Kenarsari, S.D.; Yang, D.; Jiang, G.; Zhang, S.; Wang, J.; Russell, A.G.; Wei, Q.; Fan, M. Review of recent advances in carbon dioxide separation and capture. *RSC Adv.* **2013**, *3*, 22739–22773. [CrossRef]
56. Klinthong, W.; Yang, Y.-H.; Huang, C.-H.; Tan, C.-S. A Review: Microalgae and Their Applications in CO_2 Capture and Renewable Energy. *Aerosol Air Qual. Res.* **2015**, *15*, 712–742.
57. Yang, M.; Song, Y.; Jiang, L.; Zhao, Y.; Ruan, X.; Zhang, Y.; Wang, S. Hydrate-based technology for CO_2 capture from fossil fuel power plants. *Appl. Energy* **2014**, *116*, 26–40. [CrossRef]
58. Zheng, J.; Chong, Z.R.; Qureshi, M.F.; Linga, P. Carbon Dioxide Sequestration via Gas Hydrates: A Potential Pathway toward Decarbonization. *Energy Fuels* **2020**. [CrossRef]
59. Matsuo, S.; Umeda, H.; Takeya, S.; Fujita, T.A. Feasibility Study on Hydrate-Based Technology for Transporting CO_2 from Industrial to Agricultural Areas. *Energies* **2017**, *10*, 728. [CrossRef]
60. International Energy Agency, CCUS in Clean Energy Transitions. Available online: https://www.iea.org/reports/ccus-in-clean-energy-transitions/a-new-era-for-ccus (accessed on 18 March 2021).
61. Svante. Capturing Carbon Economically, Today. Available online: https://svanteinc.com/carbon-capture-technology/ (accessed on 18 March 2021).
62. Air Products. Carbon Capture. Available online: https://www.airproducts.com/company/innovation/carbon-capture#/ (accessed on 18 March 2021).
63. Global CCS Institute. Facilities Database. Available online: https://co2re.co/FacilityData (accessed on 4 August 2020).

64. Townsend, A.; Gillespie, A. *Scalling Up the CCS Market to Deliver Net-Zero Emissions*; Global CCS Institute: Docklands, Australia, 2020; Available online: https://www.globalccsinstitute.com/wp-content/uploads/2020/04/Thought-Leadership-Scaling-up-the-CCS-Market-to-Deliver-Net-Zero-Emissions-Digital-6.pdf (accessed on 18 March 2021).
65. Budinis, S.; Krevor, S.; Dowell, N.M.; Brandon, N.; Hawkes, A. An assessment of CCS costs, barriers and potential. *Energy Strategy Rev.* **2018**, *22*, 61–81. [CrossRef]
66. Karayannis, V.; Charalampides, G.; Lakioti, E. Socio-economic Aspects of CCS Technologies. *Procedia Econ. Financ.* **2014**, *14*, 295–302. [CrossRef]
67. Stigson, P.; Hansson, A.; Lind, M. Obstacles for CCS deployment: An analysis of discrepancies of perceptions. *Mitig. Adapt. Strateg. Glob. Chang.* **2012**, *17*, 601–619. [CrossRef]
68. Ranjan, M.; Herzog, H.J. Feasibility of air capture. *Energy Procedia* **2011**, *4*, 2869–2876. [CrossRef]
69. Gutknecht, V. Awesome Extractors. Available online: https://mag.ebmpapst.com/en/industries/refrigeration-ventilation/awesome-extractors_12472/ (accessed on 6 August 2020).
70. Sandalow, D.; Friedmann, J.; McCormick, C.; McCoy, S. Direct Air Capture of Carbon Dioxide. 2018. Available online: https://www.globalccsinstitute.com/wp-content/uploads/2020/06/JF_ICEF_DAC_Roadmap-20181207-1.pdf (accessed on 18 March 2021).
71. Mazzotti, M.; Baciocchi, R.; Desmond, M.J.; Socolow, R.H. Direct air capture of CO_2 with chemicals: Optimization of a two-loop hydroxide carbonate system using a countercurrent air-liquid contactor. *Clim. Chang.* **2013**, *118*, 119–135. [CrossRef]
72. Ocean Studies Board and National Academies of Sciences Engineering, and Medicine. *Negative Emissions Technologies and Reliable Sequestration: A Research Agenda*; The National Academies Press: Washington, DC, USA, 2019.
73. Sinha, A.; Realff, M.J. A parametric study of the techno-economics of direct CO_2 air capture systems using solid adsorbents. *AIChE J.* **2019**, *65*, e16607. [CrossRef]
74. Keith, D.W.; Holmes, G.; St. Angelo, D.; Heidel, K. A Process for Capturing CO_2 from the Atmosphere. *Joule* **2018**, *2*, 1573–1594. [CrossRef]
75. Carbon Engineering Ltd. Direct Air Capture. Available online: https://carbonengineering.com/our-technology/ (accessed on 18 March 2021).
76. Beuttler, C.; Charles, L.; Wurzbacher, J. The Role of Direct Air Capture in Mitigation of Anthropogenic Greenhouse Gas Emissions. *Front. Clim.* **2019**, *1*, 10. [CrossRef]
77. International Energy Agency. *Direct Air Capture*; International Energy Agency: Paris, France, 2020. Available online: https://www.iea.org/reports/direct-air-capture (accessed on 9 April 2021).
78. Creutzig, F.; Breyer, C.; Hilaire, J.; Minx, J.; Peters, G.; Socolow, R. The mutual dependence of negative emission technologies and energy systems. *Energy Environ. Sci.* **2019**, *12*, 1805–1817. [CrossRef]
79. Fasihi, M.; Efimova, O.; Breyer, C. Techno-economic assessment of CO_2 direct air capture plants. *J. Clean. Prod.* **2019**, *224*, 957–980. [CrossRef]
80. Sandalow, D.; Friedmann, J.; Aines, R.; McCormick, C.; McCoy, S.; Stolaroff, J. *Industrial Heat Decarbonization Roadmap*; ICEF–Innovation for Coal Earth Forum: Tokyo, Japan, 2019.

energies

MDPI

Review

CO_2 Capture, Use, and Storage in the Cement Industry: State of the Art and Expectations

Marta G. Plaza *, Sergio Martínez and Fernando Rubiera

Instituto de Ciencia y Tecnología del Carbono, INCAR-CSIC, C/Francisco Pintado Fe 26, 33011 Oviedo, Spain; sergiomg691@gmail.com (S.M.); frubiera@incar.csic.es (F.R.)

* Correspondence: m.g.plaza@incar.csic.es

Received: 20 August 2020; Accepted: 21 October 2020; Published: 30 October 2020

Abstract: The implementation of carbon capture, use, and storage in the cement industry is a necessity, not an option, if the climate targets are to be met. Although no capture technology has reached commercial scale demonstration in the cement sector yet, much progress has been made in the last decade. This work intends to provide a general overview of the CO_2 capture technologies that have been evaluated so far in the cement industry at the pilot scale, and also about the current plans for future commercial demonstration.

Keywords: CO_2 capture; cement; post-combustion; absorption; membranes; adsorption; calcium looping; oxyfuel; direct separation

1. Introduction

The global production of cement reached 4.2 Gt in 2019 [1]. The main producer is China (55%), followed in the far distance by India (8%) [2]. The market is expected to grow in the forthcoming decades up to 12–23% by 2050, as a result of the increasing global population and urbanization trends, coupled with infrastructure development needs [3].

Figure 1 represents a state-of-the art cement manufacturing process. Limestone (or other calcium carbonate source) is ground with clay and other minor components, and fed to the preheater, where the raw meal countercurrently contacts the hot kiln exhaust gases in a series of vertical cyclones, being heated up to approximately 900 °C. At the bottom of the preheater, in the calciner, limestone decomposes to form calcium oxide, releasing CO_2 (see Reaction 1). This is referred to as process emissions, because they are inherent to the clinker manufacturing process. Most of the CO_2 is normally released from the limestone raw material in the preheater and the calciner [4]. The precalcined meal is heated up to 1450 °C in the rotary kiln, where the calcination is completed, and the calcium oxide reacts with silica, alumina, and iron oxides, to form the calcium silicates, aluminates, and ferrites that constitute the clinker. At the kiln outlet, the clinker is rapidly cooled in the grate cooler using incoming combustion air. Finally, the cooled clinker is mixed with gypsum and ground into a powder, to produce Portland cement, or with additional components—such as slag, fly ash, or limestone—that substitute part of the clinker, to produce blended cement. Each cement product has its own unique composition of raw materials, which depends strongly on the location of the cement plant and the availability of the raw materials.

$$CaCO_3CaO + CO_2 \quad (1)$$

Due to its size and the inherent characteristics of its production process, the cement sector is one of the main sources of anthropogenic CO_2, accounting for 8% of global emissions [5]. Up to 0.95 t of CO_2 are released per t of cement produced. The actual carbon footprint depends on the ratio of clinker to cement, the manufacturing process (dry or wet method), the level of heat recovery, the fuel used, the moisture content of the raw materials, and the capacity of the plant, among other factors. Process emissions account for approximately 65% of the direct CO_2 emissions, whereas fuel combustion is responsible for the remainder [6]. Total CO_2 emissions have increased by 200% since 1990, mainly driven by the growth in cement demand, despite the decrease in the global average clinker ratio [7], the significant improvement on energy efficiency, and the increasing rate of alternative fuels used [3]. However, coal still remains the dominant fuel, with a share of 65% of the total fuel consumption [6]. On the other hand, the net climate impact of concrete products needs to account for the fact that 11–43% of the initial process emissions may be offset by gradual reabsorption of atmospheric CO_2 by the exothermic carbonation of hydrated cement [8].

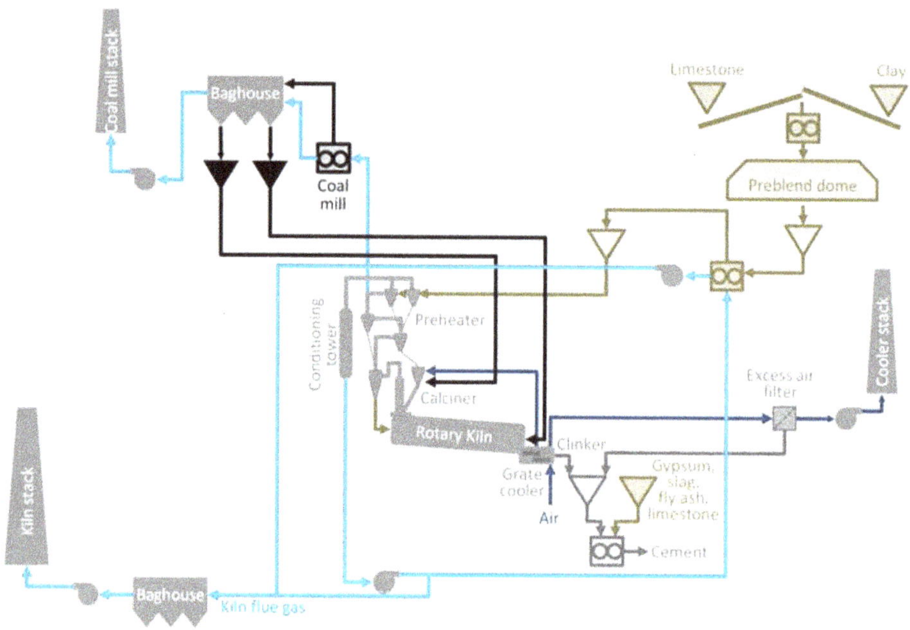

Figure 1. Scheme of a state-of-the art cement manufacturing process.

The Paris Agreement, ratified by 189 of the Parties, aims to limit global warming to well-below 2 °C above pre-industrial levels and to pursue efforts to restrain it to 1.5 °C. Limiting warming to 1.5 °C implies reaching net zero CO_2 emissions globally around 2050 [9]. Therefore, it is necessary to decarbonise the cement industry at a global scale.

All regions have improved the energy efficiency and reduced the carbon intensity of their cement industries. China, India, and the European Union, which are the three main cement producers, have adopted policy measures to reduce the carbon footprint of the cement industry, through emissions trading schemes, energy efficiency targets, replacement of old plants, new cement standards, and circular economy. Private-led initiatives are also gaining momentum. In 2015, 18 cement companies established the shared ambition to reduce their CO_2 emissions by 20–25% compared to business as usual by 2030, which represents a mitigation effort of about 1 Gt CO_2 [3]. The Global Cement and Concrete Association (GCCA) includes worldwide members that sum up to 50% global cement production capacity. In 2019, GCCA launched Innovandi, the Global Cement and Concrete Research Network, which intends to research areas such as the impact of co-processing, the efficiency of clinker production, the implementation of Carbon Capture Use and Storage (CCUS) technologies, the impact of clinker substitutes and alternative binders in concrete, low carbon concrete technologies, and re-carbonation [10]. Thirty companies from the cement and concrete sectors, have already committed to the initiative [11]. The European Cement Association (CEMBUREAU) has publicly stated its ambition to reach carbon neutrality along the value chain by 2050 [12]. HeidelbergCement has committed itself to reduce its specific net CO_2 emissions per ton of cement produced by 30% compared to 1990 levels by 2030, and has a vision to reach carbon neutral concrete by 2050 at the latest; this is the first cement company in the world to receive approval for science-based CO_2 reduction targets [13], and the only cement manufacturer that has been included in the CDP's Climate Change A list 2019 [14]. It aims to attain net zero carbon footprint by augmenting the percentage of CO_2-neutral feed materials and combustibles, arriving at lower clinker cement classes, and by CCUS. Dalmia cement has also committed to become carbon neutral by 2040. Other companies are heading the same way.

In the International Energy Agency's (IEA) 2 °C scenario (2DS), which is consistent with at least a 50% probability of limiting the average global temperature increase to 2 °C by 2100, the direct emissions of the cement sector are reduced by 24% by 2050. This implies a cumulative reduction of emissions of 7.7 Gt CO_2 compared to the IEA's Reference Technology Scenario (RTS), which considers the commitments by countries to limit carbon emissions, including the nationally determined contributions (NDC) pledged under the Paris Agreement, which would result in an average temperature increase of 2.7 °C [3]. With increasing climate ambition, the mitigation effort grows larger: in order to have a 50% probability of limiting the average global temperature increase to 1.75 °C, (IEA's Beyond 2 °C Scenario, B2DS) additional 3.2 Gt CO_2 of cumulative CO_2 emissions reduction would be necessary compared to 2DS [3]. Obviously, if the most ambitious climate objective of the Paris Agreement—i.e., limiting the global average temperature increase to 1.5 °C—is to be pursued, the mitigation effort would need to increase accordingly. However, according to CEMBUREAU's carbon neutrality roadmap, reaching net zero emissions along the cement and concrete value chain is achievable by 2050 [12].

There is a series of mitigation measures that can contribute to the decarbonization of the cement industry: improving energy and materials efficiency (on a life cycle approach); switching to less carbon intensive fuels; reducing the clinker content in the cement by partly replacing it with cementitious materials with lower carbon footprint; developing new, innovative and clean production technologies, including excess heat recovery (EHR) to power generation, integration of renewable power generation, CCUS; and improving transport efficiency [3].

As shown in Figure 2, in the 2DS, the largest cumulative direct CO_2 emissions reduction by 2050, compared to RTS, comes from the reduction in the clinker ratio, followed by CCUS, development of clean production technologies, fuel substitution, and thermal efficiency [3].

Broadly, CCUS prevents CO_2 from being released into the atmosphere by capturing it, and either using it, or injecting it in geological formations for permanent storage. Under the B2DS, the amount of CO_2 captured annually by 2050 more than doubles that of the 2DS (see Figure 3), and the share of total direct CO_2 emissions cumulatively captured increases from 25% to 63%. To achieve carbon neutrality by 2050, the role of CCUS would be even higher, as this is necessary to achieve deep levels of decarbonization, given the large share of process emissions in the cement sector. The cement industry undoubtedly needs to deploy CCUS technologies. Although there are not yet CCUS facilities operating in the cement sector at large scale, there are CCUS facilities currently operating at the power and industrial sectors with a total annual capacity close to 40 Mt CO_2 [15].

This work intends to review the current state of the art of the CCUS technologies in the cement sector, including pilot scale demonstration studies and announced large scale projects that could start operation in the near term.

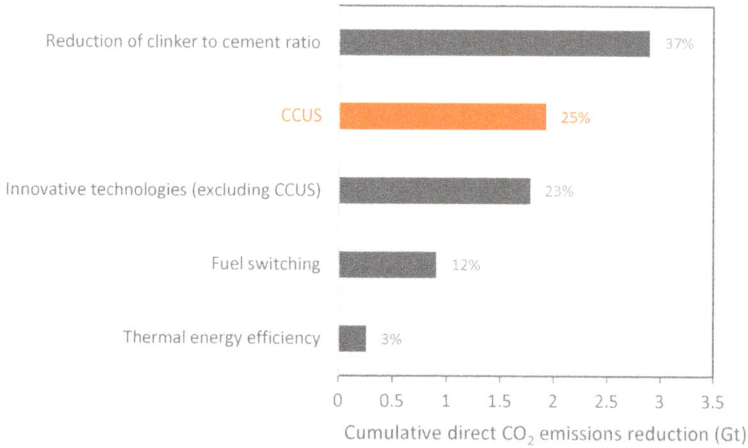

Figure 2. Global cumulative CO_2 emissions reductions by 2050 (2DS vs. RTS) (adapted from [3]).

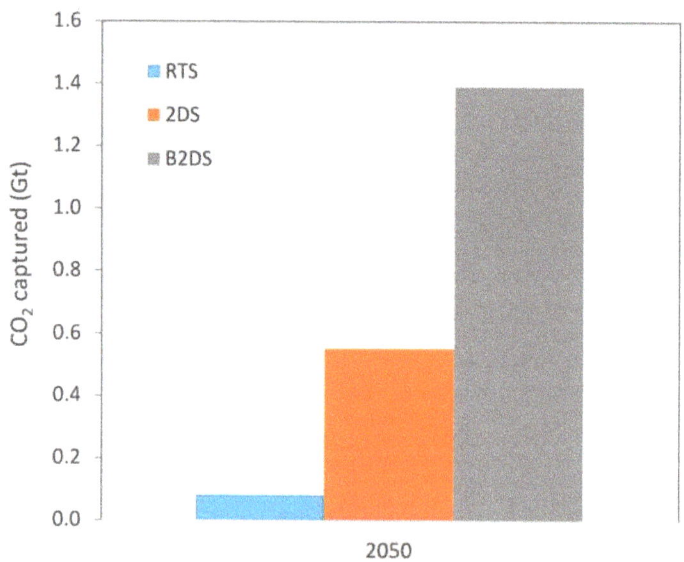

Figure 3. Global direct CO_2 emissions from the cement sectors captured annually according to different IEA scenarios (data taken from [3]).

2. State of the Art of CO_2 Capture in the Cement Industry

In the Intergovernmental Panel on Climate Change's (IPCC) Special report on CCS, launched in 2005, four different types of CO_2 capture technologies were distinguished: post-combustion, pre-combustion, oxyfuel combustion, and industrial separation [16]. This classification differentiates industrial CO_2 capture from that carried out in the power sector. However, the separation technologies that can be used to capture CO_2 in industry can have many similarities with those of the power sector, depending on the specific process considered. In the case of the cement industry, CO_2 capture can be accomplished using post-combustion and oxyfuel combustion technologies.

Pre-combustion capture technologies would have limited mitigation potential in the cement sector, as they could deal only with the energy-related CO_2 emissions [17]. They might be of use in new cement plants integrated with gasification technologies to produce syngas or H_2 fuel [18]. However, hydrogen flames have a relatively low emissive power, which makes them unsuitable for clinker manufacturing in conventional kilns [19]. New and more efficient hydrogen burners and cement kiln lines would be required.

A third type of CO_2 capture technology with great promise to be implemented in the cement sector is direct capture.

The state of the art of these groups of technologies in the cement sector will be discussed in the following sections.

2.1. State of the Art of Post-Combustion CO_2 Capture Technologies in the Cement Sector

A priori, conventional kilns of existing and new cement plants could be retrofitted with post-combustion CO_2 capture technologies relatively easily, without substantially modifying the cement manufacturing process. Only the energy management strategies and the start-up and shut-down procedures would be affected [19,20]. The inherent characteristics of the sector, the fact that actual kilns still have several life years ahead (expected 30–50 years lifetime), and that new kilns adapted to legal requirements are not thought to be built in the next years, reinforce the potential of post-combustion CO_2 capture technologies [21].

The kiln off-gas holds a temperature in the range of 85–105 °C to 150–180 °C, depending on the manufacturing process, and is at near atmospheric pressure [22]. These conditions are not ideal for the point of view of the separation process. On the other hand, the kiln-off gas presents a higher carbon dioxide concentration (up to 30% by volume after the preheater [19]) than other flue gas streams, including those of coal-fired power plants, which facilitates the capture step.

There are different post-combustion technologies that can be used to capture the CO_2 from the kiln-off gas: chemical absorption, membranes, and sorption with solids. A brief review of the state of the art of each of them is presented below.

2.1.1. Chemical Absorption

SkyMine® Process

The largest demonstration of CCUS in the cement sector to date, started to operate in September 2015, at Capitol Aggregates' San Antonio cement plant in Texas, United States of America, with a capacity of over 75,000 t CO_2/y. The pilot plant was designed to make use of the SkyMine® process to capture 90% of the CO_2 from a slipstream of the cement plant, which represented approximately 15% of its total CO_2 emissions. The SkyMine® process produces marketable by-products, such as baking soda, hydrochloric acid, and bleach (see Figure 4), with a lesser CO_2 footprint than the conventional production processes, which contributes to additional CO_2 savings by product displacement (up to a total benefit of 224,110 t CO_2/y) [23]. The kiln-off gas, with an annual average composition of 29.58 vol % CO_2, 11.20 vol %H_2O, 10.22 vol % O_2, 0.04 vol % SO_2, 0.01 vol % NO, and 0.05 vol % CO, is cooled from approximately 125 °C to 35 °C. A diluted solution of NaOH is used to knock out SO_2 and NO_x. 60% of heavy metals (mainly Hg) are removed with the condensate, which is filtered with activated carbon to recycle water to the process. Conditioned flue gas is then fed to a multicolumn chemical absorption system, where a concentrated NaOH solution reacts countercurrently with the CO_2 from the flue gas in two packed absorbers working in parallel to form Na_2CO_3 (Reaction 2). Then, the saturated Na_2CO_3 solution is fed to bubble column reactors where it reacts with pressurized flue gas to yield $NaHCO_3$ crystals (Reaction 3). The resultant slurry is separated in a centrifuge into solids, which are sent to the dryer, and liquids, which are further processed to recycle the concentrated solution to the bubble column reactor and excess water to the process. The NaOH solution is generated through electrolysis of NaCl brine in a membrane cell that produces HCl as byproduct [23,24].

$$2NaOH_{(aq)} + CO_{2\,(g)} Na_2CO_{3(aq)} + H_2O_{(l)} \tag{2}$$

$$Na_2CO_{3(aq)} + H_2O_{(l)} + CO_{2\,(g)} 2NaHCO_{3(s)} \tag{3}$$

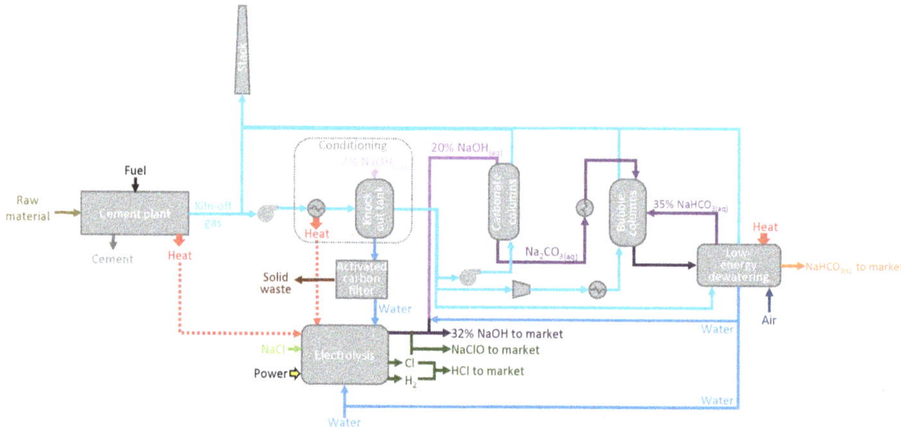

Figure 4. Simplified diagram flow of the SkyMine® process integrated in San Antonio cement plant (adapted from [23,24]).

According to available data, electrolysis is responsible for 87% of the total power requirement of the overall process, which is 21.12 MW_e. The specific power consumption results 8.3 GJ_e/t CO_2 captured, or 2.8 GJ_e/t CO_2 avoided, if the additional CO_2 savings achieved by the displacement of more energy-intensive commodities in the market are considered. The total installed cost of the project was estimated to be \$125 M [23].

Amine Scrubbing

Chemical absorption with amine solutions has been extensively used in industry since 1930 [25], and has reached commercial scale demonstration in coal fired power plants [26,27].

Although specific process details would depend upon the particular facility, a generic process has three main steps: (i) flue gas pretreatment in a Direct Contact Cooler (DCC), with SO_2, NO_x, and dust particle removal; (ii) countercurrent contact of pretreated flue gas with an aqueous amine solution in an absorber column, where CO_2 reacts with the amine at 40–60 °C and atmospheric pressure, producing a decarbonized gas stream that is vented to the stack; (iii) regeneration of the spent solvent in the stripper column, at 100–120 °C and 1.5–2 atm (the operational values change slightly depending on the solvent used [28]), where a high purity CO_2 stream is recovered, while the lean solvent is sent back to the absorber column, closing the loop [29].

Aqueous solutions of alkanolamines, like monoethanolamine (MEA) and diethanolamine (DEA), were traditionally used because of their rapid reaction rates and low cost [30]. The development of second generation solvents and the optimization of the technology has led to substantial energy savings, up to 60% in the power sector [15].

The main differences between the flue gas of a power plant and that of a cement factory, are its temperature, composition, and different size distribution of their particulate matter. The higher temperature of kiln-off gases would cause thermal degradation of the amines and greater losses through evaporation [31], and hence requires further cooling. The higher CO_2 partial pressure in the cement flue gas is a priori advantageous, although it can also be a challenge for the CO_2 absorber, as more heat of absorption is released for a smaller volume of flue gas, and thus, the CO_2 absorber temperature is likely to increase, shifting the equilibrium in a less favorable direction [32].

SO_2 reacts irreversibly with amines to produce corrosive salts, while NO_X can produce nitric acid resulting in additional corrosion and amine degradation [19]. To avoid the aforementioned problems, it is necessary to remove the SO_2 and NO_2 levels below 10 and 20 ppm, respectively, prior to the absorber [33]. This pretreatment significantly increases the capital and operating cost of the overall process.

The main disadvantage of amine scrubbing is its high energy demand, mostly driven by the solvent regeneration, which contributes with 50–80% of the total energy requirements [34]. Moreover, in a cement plant there is no source of low pressure steam available to regenerate the solvent: at least 2 GJ/t CO_2 captured would need to be provided via a combined heat and power facility (CHP), or through waste heat recovery [20]. Furthermore, in modern cement plants, a significant fraction of waste heat is already used to dry the raw materials and preheat the raw meal. According with some simulations, no more than 15% of the additional thermal energy needs can be recovered from the cement kiln [35]. However, the actual amount of surplus waste heat is site dependent; this is strongly influenced by the moisture content of the raw materials. Calculations conducted for the Norcem's Brevik cement plant, in Norway, where the raw materials are relatively dry, indicate that using only waste heat, approximately 40% of the total CO_2 emitted by the plant could be captured, which correspond to 0.4 Mt CO_2/y [22]. More extensive heat integration (and costly) could lead to even higher capture rates, up to 85% of total emissions, using only waste heat [36].

The cost of the technology is dependent on the site location, the specific technology used, the steam source, the fuel price, and plant specific characteristic. The estimated cost for the retrofit of a cement plant with the reference MEA-based absorption technology is 80 €/t CO_2 avoided [37].

Although amine scrubbing is considered the benchmark technology for CO_2 capture in the cement industry, it is yet to be demonstrated at large scale in a cement plant. Norway's Longship project might be the first to achieve that milestone, being the first of its kind in the world. Aker Solutions' Advanced Carbon Capture (ACC^{TM}) technology (see Figure 5) and its S26 amine solvent, will be used to capture 0.4 Mt/y of the CO_2 emitted by the Norcem's cement factory located in Brevik, Norway, making use of available waste heat. The captured CO_2 will be liquefied and temporarily stored at Brevik facilities, and then shipped to an onshore terminal at Øyrgarden, on the Norwegian west coast, from where it will be transported by pipeline to an offshore storage location under the North Sea within the Northern Lights transport and storage project. The Norwegian Government has recently approved the final investment decision [38]. Operation is expected by 2023 [39]. Total investment in Longship project is estimated to be NOK 25 billion, although this also covers CO_2 capture at Fortum Oslo Varme's waste incineration facility [38].

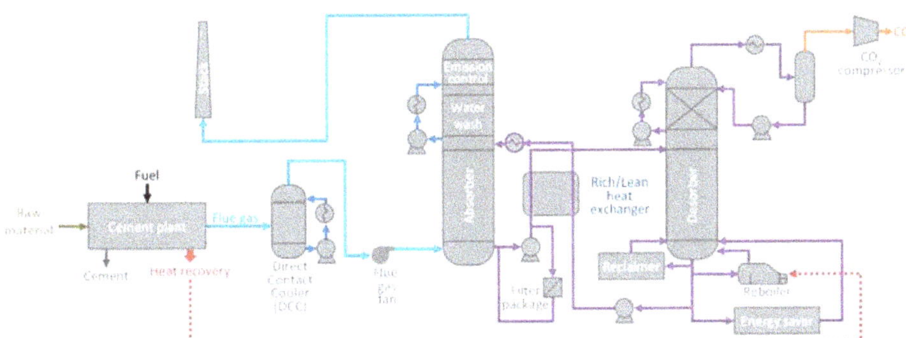

Figure 5. Generic flow sheet of an ACC^{TM} process integrated in a cement plant.

Aker Solutions has extensively tested its ACC^{TM} technology making use of its advanced amine solvent S26 at pilot scale in Brevik factory for 18 months [40]. The Mobile Test Unit (MTU), installed at Brevik in April 2014, has a flue gas capacity up to 1000 Nm^3/h, and consists of an absorber with a diameter of 0.4 m and a packing height up to 18 m, a desorber with a diameter of 0.32 m and a packing height of 8 m, and a proprietary anti-mist design. Typically, a slipstream of 450 Nm^3/h of the cement kiln flue gas was taken just before the stack, downstream of the selective non-catalytic reduction (SNCR), electrostatic precipitator (ESP), spray dryer flue gas desulfurization (FGD), and baghouse filter, with the following composition: 7.5 vol % O_2, 17.8 vol % CO_2, 18.2 vol % H_2O, 0–130 mg/Nm^3 SO_2, 180–250 mg/Nm^3 NO_x and 5–10 mg/Nm^3 dust. This corresponds to a CO_2 removal rate of 140–150 kg/h which is the full capacity of the reboiler [32]. SO_2, NO_x, and dust particles were knocked out from the flue gas with the condensate [22]. During field testing, which included over 5500 h, stable operation on flue gas from cement kiln was demonstrated, with 90% CO_2 capture, low amine consumption (<0.15 kg/t CO_2), low build-up of degradation products (heat stable salts mainly formed by slip of SO_2 through pre-scrubber), with low nitrosamine formation, low corrosiveness, and low emissions. The energy demand, without heat integration, for a 90% CO_2 removal rate is approximately 3 GJ/t CO_2, or 2.7 GJ/t CO_2 with the ACC™ Energy Saver, and this can be entirely satisfied using the available low grade heat from the cement plant. Aker Solution's ACC™ technology has been scaled-up and the S26 solvent qualified at Technology Centre Mongstad [36]. The technology has currently a technology readiness level (TRL) of 8 [39].

To date, the largest demonstration of CO_2 capture using amine absorption technology in the cement industry, with a capacity of 50,000 t CO_2/y, is located in Anhui Conch's Baimashan plant in the city of Wuhu, in the Ahnui province, China [41]. The pilot facility, which started to operate in 2018, captures 3% of the total CO_2 emissions of the cement factory. The CO_2 produced, with a purity of 99.99%, is transported by trucks and sold to industrial customers [42]. However, due to the limited local market for CO_2, the company has no plans to expand.

Dalmia Cement announced in 2019 its intention to build a large scale CCUS facility at one of its cement plants in Tamil Nadu, India, with a capacity of 0.5 Mt CO_2/y [43]. The facility will make use of Carbon Clean's CDRMax® technology, which combines the use of a proprietary solvent (amine promoted buffer salts, APBS) with novel heat integration, to provide CO_2 capture at an estimated cost of \$40/t CO_2. The technology can be flexed to manage CO_2 concentrations in the source gas from 3% to 25% and to produce CO_2 with purities between 95–99.9% [44]. The partnership will explore possible uses for the CO_2 captured, including direct sales to other industries, or chemical manufacturing. No completion date or budget has been disclosed yet.

Chilled Ammonia Process

The chilled ammonia process (CAP) makes use of an ammonium aqueous solution to absorb CO_2 from the flue gases at ambient pressure and low temperature. This entails the use of low grade heat. Moreover, the solvent is not affected by oxygen or acidic trace components present in the incoming flue gas. The main chemical reactions involved in CAP are listed below [45]:

$$CO_{2\,(g)} \leftrightarrow CO_{2(aq)} \tag{4}$$

$$2NH_{3(aq)} + H_2O_{(l)} + CO_{2\,(aq)} \leftrightarrow (NH_4)_2CO_{3(aq)} \tag{5}$$

$$(NH_4)_2CO_{3(aq)} + H_2O_{(l)} + CO_{2\,(aq)} \leftrightarrow 2(NH_4)HCO_{3(aq)} \tag{6}$$

$$(NH_4)_2CO_{3(aq)} \leftrightarrow NH_2CO_2NH_{4(aq)} + H_2O_{(l)} \tag{7}$$

Reactions 4–7 are reversible and the direction depends on pressure, temperature and concentration.

A simplified flow diagram of CAP process is shown in Figure 6. The flue gas is first conditioned in a combined DCC and SO_2 absorber, where the flue gas is cooled and SO_2 is scrubbed with ammonia. Liquid ammonium sulphate $(NH_4)_2SO_4$ is formed as a by-product, which can be commercialized as a fertilizer. Optionally, ammonia can be recovered in a dedicated recovery unit, leading to a gypsum by-product [45]. The conditioned flue gas is then sent to an absorption column where CO_2 is removed by an ammonia solution at temperatures around 12–13 °C. To reduce the ammonia loses with the decarbonized flue gas, a water wash section is installed at the top of the absorber; ammonia is recovered from the wash water in a dedicated stripper and recycled to the CO_2 adsorber. To further reduce the emissions of ammonia to the atmosphere, an acid wash section fed with H_2SO_4 is installed between the water wash and the stack. The spent solvent is regenerated in the CO_2 desorber at around 25 bar [46].

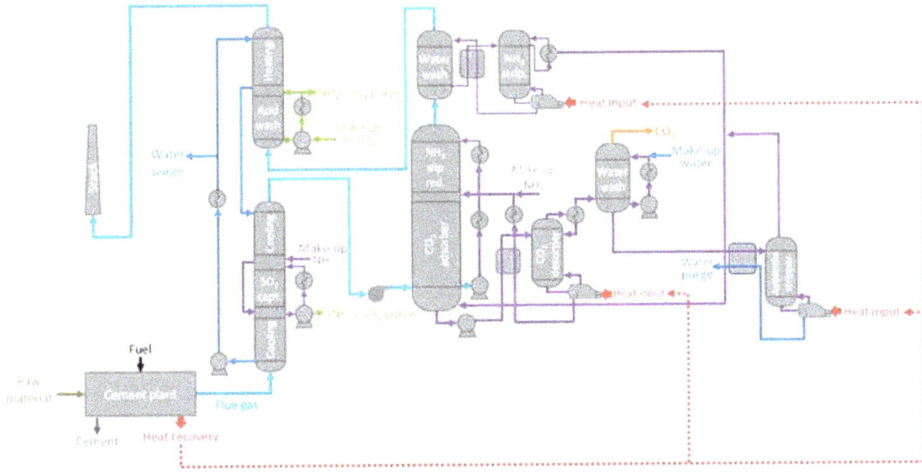

Figure 6. Process flow sheet of a CAP process integrated in a cement plant (adapted from [46]).

Within the European funded CEMCAP project (2015–2018), TRL of GE's CAP process for cement plants was advanced to TRL 6. Three experimental campaigns were carried out at a 1 t CO_2/d pilot plant of GE's Technology Center in Växjö, Sweden. The first campaign evaluated the effect of high CO_2 concentration in the absorber; the second campaign studied the combined DCC and SO_2 absorber, which reduces the SO_2 content below 1 ppm; and the last campaign investigated the flue gas water wash, which reduces the NH_3 slip from 1% to below 200 ppm [47]. Tests proved that operating conditions could be adapted to work under cement plant conditions, with CO_2 recovery rates of 90%, using a similar design than that of power plants. Moreover, the higher CO_2 concentrations of cement plant flue gases allowed enhancing the NH_3 absorber design, reducing its height and complexity. CAP is ready to be demonstrated at a scale of 100,000 t CO_2 captured/y in a cement plant [48]. For a standard 1 Mt clinker/y plant, the expected steam consumption for a 88% capture rate is 2.37 GJ steam/kg CO_2 [47]. The specific primary energy consumption per CO_2 avoided (SPECCA) of the CAP process for a 90% CO_2 capture rate is 3.75 GJ/t CO_2 captured [46], while the estimated cost of CO_2 avoided is around 66.2 €/t CO_2 [37].

2.1.2. Membrane Separation

Gas separation membranes are based in physical and/or chemical interactions between the different gas components and the membrane material: one component passes through the membrane faster than another, being separated by the pressure gradient and the different diffusivities of the molecules. Membranes can be ceramic, polymeric (the most common), or hybrid. Selectivity is provided by the membranes themselves, and permeation rates would depend on parameters such as the relative molecular sizes, or their diffusion coefficient in the membrane. The driving force for the permeation across the membrane is the difference in the partial pressure of the components at both ends.

Membranes can be designed for a wide variety of roles, including carbon capture in the cement industry [19]. The higher concentration of CO_2 in cement flue gas compared to other sources, is advantageous for all separation technologies, and especially for membranes, because of the larger partial pressure driving force [49].

Membrane based separation selectivity is low, and only a fraction of the inlet CO_2 can be captured. In addition, the CO_2 purity is also limited, so usually multistage operation is required to fulfill the desired product standards. Membranes could theoretically reach capture rates larger than 80%, although only up to 60–70% recovery has been demonstrated at laboratory scale [17]. Purity obtained is about 90% (in a typical two-stage process), but it can be increased over 95% through new stages addition or by adding a low-temperature liquefaction unit [50]. Moreover, membranes are usually sensitive to sulfur traces, and polymeric membranes are mostly intolerant to high temperatures. Nevertheless, membrane systems do not require regeneration (although they require to be operated under pressurized conditions, or vacuum, which imply a high energy demand), and their overall footprint is lower than that of other technologies [17].

Membrane separation was demonstrated at pilot scale in Norcem CO_2 capture project (2013–2017), in Brevik cement plant, in Norway. The pilot facility, shown in Figure 7, was designed and constructed by Yodfat Engineers and operated by the Norwegian University of Science and Technology (NTNU) and DNV GL. This consisted of a one stage membrane module with 12 cassettes, with 2 flat sheets of polyvinylamine (PVAm) based fixed-site carrier (FSC) membranes per cassette, made by NTNU, with a total membrane area of approximately 1.5 m^2. Initial testing, carried out in 2014 with a slipstream of the plant flue gas downstream the SNCR, ESP, FGD, and baghouse filter, showed that it was difficult to achieve a stable and high performance of the membrane system. A CO_2 purity up to 72 vol % was achieved for short periods of time, when all process parameters were well controlled in the single stage FSC membrane system. The membranes withstood extended exposure to cement flue gas without showing damage, with promising recoveries in the range of 60–70% [22]. Nevertheless, the membrane efficiency in the plat-and-frame module was low, and suffered water condensation/corrosion issues.

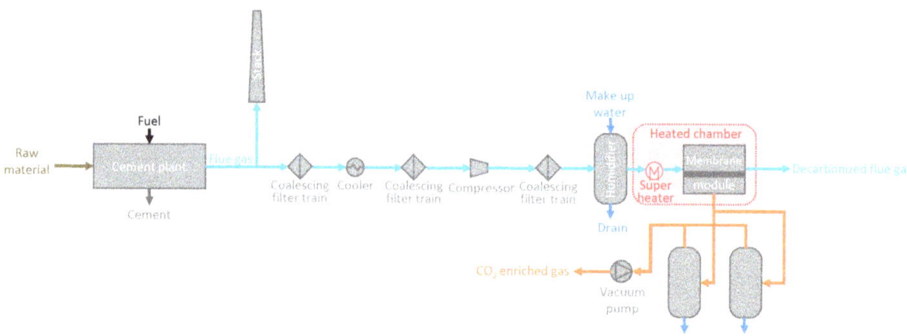

Figure 7. Simplified flow diagram of the pilot membrane unit tested at Brevik cement plant (adapted from [51]).

In the MemCCC project, Air Products' commercial hollow fiber membrane modules were coated at NTNU with the PVAm based FSC membrane (up to 18 m^2), and tested at Brevik plant for 9 months at different conditions (including high NO_x and SO_2 concentrations) showing stable performance, and up to 70 vol % purity in a single stage. The techno-economic analysis for 80% CO_2 recovery and a purity of 95 vol %, showed a power consumption of 1.20 GJ_e/t CO_2 captured, and an estimated cost of 46.54–48.49 €/t CO_2 captured [51]. These results could be enhanced adding new stages or a low-temperature liquefaction unit [52].

In membrane-assisted CO_2 liquefaction (MAL) technology, the bulk separation of CO_2 is carried out by permeation through a membrane module, resulting in a permeate with a moderate purity, which is fed to a low temperature liquefaction unit, where it is conditioned, compressed, and cooled typically between −55 and −50 °C, leading to a high purity CO_2 liquid product and a gas stream that is vented to the stack, as shown in Figure 8.

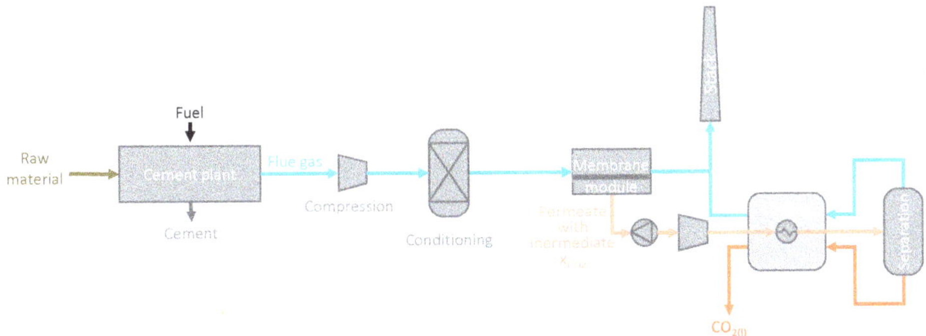

Figure 8. MAL system integrated in a cement plant (adapted from [53]).

Within CEMCAP project, MAL has been evaluated for cement production. Two polymeric membrane materials, perfluoropolymer and PEBAX-based membranes, were evaluated at bench scale at TNO, in Eindhoven, Netherlands, using CO_2/N_2 mixtures reaching high separation factors and selectivities of 20 and 45%, respectively, and a 78 mol % CO_2 concentration in the permeate side for the perfluoropolymer membranes, reaching the concentration requirements for membrane assisted CO_2 liquefaction. At the same time, CO_2 was separated and purified in a laboratory pilot plant at SINTEF Energy Research in Trondheim, Norway, with a capacity of 10–15 t/day. The low-temperature liquefaction process showed stable and robust, and it is easily scalable to a larger capacity [48]. Nevertheless, tests included only N_2/CO_2 mixtures. It is still necessary to carry out additional tests containing typical kiln-off gas impurities to establish the realistic obtainable purity of the captured CO_2. The estimated cost of retrofitting this technology to cement plants is 83.5 €/t CO_2 avoided, which is significantly higher than that of competing technologies [37].

2.1.3. Sorption with Solids

Adsorption is a surface process where the adsorbate molecules, originally present in the bulk fluid phase, tend to concentrate onto the surface of a solid adsorbent. When the phenomenon is not restricted to the surface of the solid, the term 'sorption' is preferred. Solids' sorption-based separation processes are led by selective attractions between a particular adsorbate, present in the bulk fluid mixture, and a solid sorbent. There are two principal mechanisms of gas sorption by solids: physical sorption (or physisorption) and chemical sorption (or chemisorption). These are differentiated by the nature of the interaction between the sorbate and the sorbent: intermolecular forces (van der Waals) or chemical bonds formation, respectively.

Sorption with Solids at Low Temperature

Candidate physisorbents for CO_2 capture are carbon materials, zeolites, alumino-phosphates (AlPOs), alumino-silico-phosphates (SAPOs), or metal–organic frameworks (MOFs) [54]. Chemisorbents generally include an active phase, such as amines or carbonates, supported on a porous support. Sorbent development seeks to achieve larger CO_2 adsorption capacities, faster kinetics, larger selectivity towards CO_2, milder regeneration conditions, better stability, tolerance to impurities, and lower costs.

In order to operate in a continuous basis, the adsorbent can be regenerated by decreasing the pressure, in pressure swing adsorption (PSA) processes, by increasing its temperature in temperature swing adsorption (TSA) processes, by using a purge gas in concentration swing adsorption (CSA) processes, or by a mixture of the previously mentioned methods of regeneration [55]. As the flue gas of a cement plant is at near atmospheric pressure, the adsorbent can be regenerated at subatmospheric pressures in vacuum swing adsorption (VSA) processes. Process development seeks to reduce the overall energy consumption of the capture processes, while meeting the purity and recovery constraints.

The main advantage of adsorption-based CO_2 capture processes is their negligible environmental impact, with no toxic emissions or wastes, and without corrosion problems. Moreover, adsorption presents scope to reduce the energy penalty of the benchmark technology, amine absorption, although it has a lesser development stage.

The Research Triangle Institute (RTI)'s solid sorbent technology was evaluated against Aker solution's ACC^{TM} amine technology, membrane separation, and GE's regenerative calcium cycle (RCC) within Norcem CO_2 capture project (2014–1017). RTI's technology makes use of a high capacity chemisorbent (35 wt % Polyethyleneimine, PEI, loaded on silica), which flows countercurrently with flue gas in a multi-staged fluidized moving bed (SFMB) sorber, in a TSA process, where the sorbent circulates continuously between the sorber and the sorbent regenerator (also a SFMB) (see Figure 9). This type of reactor provides efficient heat management, critical to the chemisorption process, as any temperature rise in the absorber would reduce the CO_2 capacity of the sorbent. RTI evaluated the sorbent exposure to actual cement flue gas using an automated sorbent test rig (ASTR), which was installed at the Brevik cement plant, in Norway, during the first phase of the project. A slipstream of the flue gas was cooled, and the condensate knocked out. This demonstrated an effective way to reduce the contaminants that could impact RTI's sorbent, as most of them were removed with the condensate. Tests showed that exposure to 100 ppm of SO_2 caused a 30% drop in the CO_2 capacity of the sorbent. A RTI's prototype, with a sorbent inventory of 75 kg, was operated at Brevik cement plant between September and November 2016. Steady 80–90% CO_2 capture rate was demonstrated with an energy consumption of 2.4 GJ/t CO_2 avoided (without heat recovery), which is lower than that of the ACC^{TM} process. According to the technical and economical assessments carried out, the cost of CO_2 capture, for a minimum of 85% capture rate, without heat integration, is 45.8 €/t CO_2 avoided. For a minimum of 85% capture rate, but taking advantage of the available waste heat within the cement plant, it falls to 40.7 €/t CO_2 avoided. Using all the low grade heat available, without a minimum capture rate restriction, the cost falls to 38.6 €/t CO_2 avoided [56].

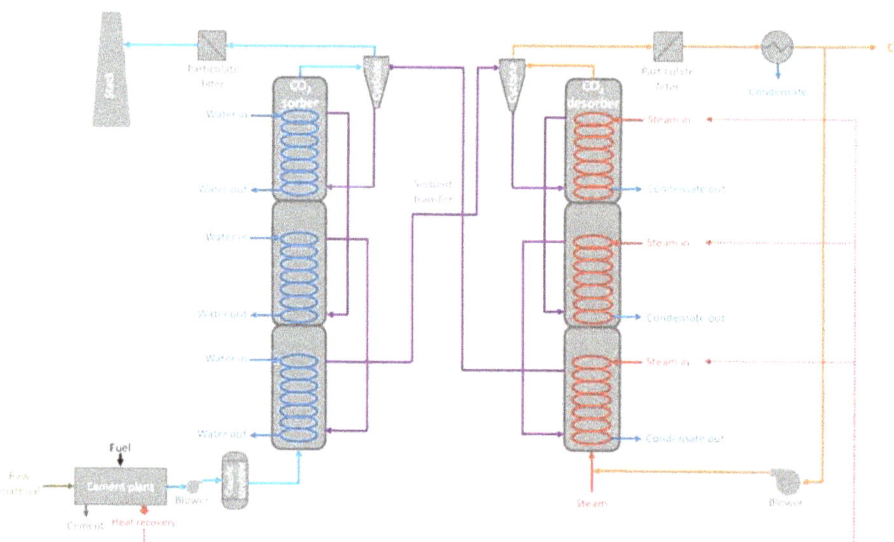

Figure 9. RTI's solid sorbent technology (adapted from [56]).

Adsorption-based technology might be close to commercial scale demonstration in the cement sector: in January 2020, Svante (formerly Inventys), LafargeHolcim, Oxy Low Carbon Ventures, and Total, launched a joint study to assess the viability and design of a commercial-scale carbon capture facility at the LafargeHolcim's Portland Cement Plant in Florence, Colorado, United States of America [57]. This initial scoping study was successfully completed in June 2020, and in September 2020 the United States Department of Energy's National Energy Technology Laboratory (DOE-NETL) awarded $1.5 million of federal funding to support the advancement of the LH CO2MENT Colorado Project. This would be the largest demonstration in the cement sector, with a capacity of up to 2 Mt of CO_2 captured annually [58]. The carbon capture facility would employ Svante' CCS technology, which is an intensified TSA process that makes use of a patented architecture of structured adsorbent (spaced sheets), and a proprietary process cycle design, the Veloxotherm^{TM} process, where a rotary adsorber is used to capture the CO_2 from the flue gas, and release it by displacement with steam (and vacuum, VTCSA), regenerating the adsorbent in a continuous manner and in a single unit (see Figure 10). Svante's process claims to be material agnostic: it can make use of different adsorbents, from carbons, to functionalized-silicas or metal–organic frameworks. The latter have high potential due to sharper temperature and pressure swing adsorption and desorption, which leads to lower parasitic energy loads, faster kinetic rates, and higher working capacity compared to conventional adsorbents, although they are still in development. As advanced adsorbents develop, cost reduction can be expected [59]. According to Svante's CEO, the abatement cost for the cement plant would be roughly $50/t CO_2. This fact, combined with the 45Q tax credit, which provides $35/t CO_2 stored through enhanced oil recovery (EOR), and the current CO_2 prices, of around $20 t CO_2, makes the project economically feasible [60].

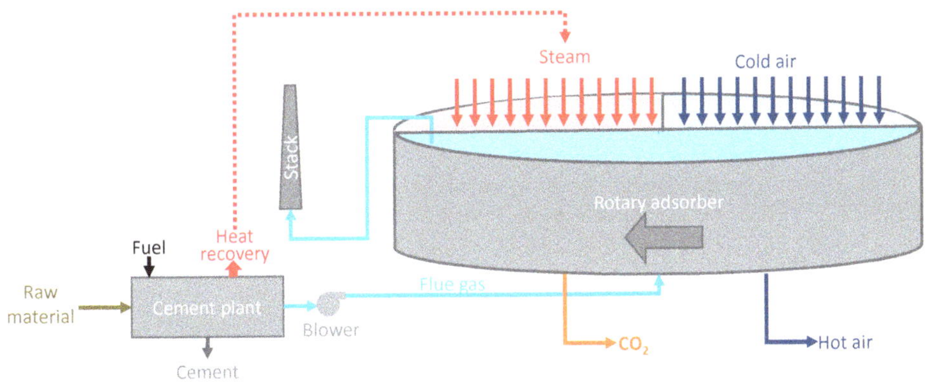

Figure 10. Simplified flow diagram of Svante's Veloxotherm^{TM} process integrated in a cement plant.

Svante's technology is being demonstrated at LafargeHolcim's cement plant in Richmond, British Columbia, Canada, at the smaller pilot scale of 1 t CO_2/d within the CO_2MENT project (2019–2022), funded by the CO_2 Capture Project (CCP), the province of British Columbia and Canada's federal government. The project was launched in partnership with Total, with the objective to build the world's first full-cycle solution to capture and reuse CO_2 from a cement plant [61]. CO_2 capture started by the end of 2019 [59]. The CO_2 captured is currently being used for synthetic fuels production and injection in concrete and fly ash.

Calcium Looping

Calcium looping (CaL) is one of the most promising CO_2 capture technologies for the cement sector. It is based on the reversible carbonation (Reaction 8), which is generally carried out in two interconnected circulating fluidized beds: a carbonator and a calciner (see Figure 11).

$$CaO + CO_2 \rightleftharpoons CaCO_3 \tag{8}$$

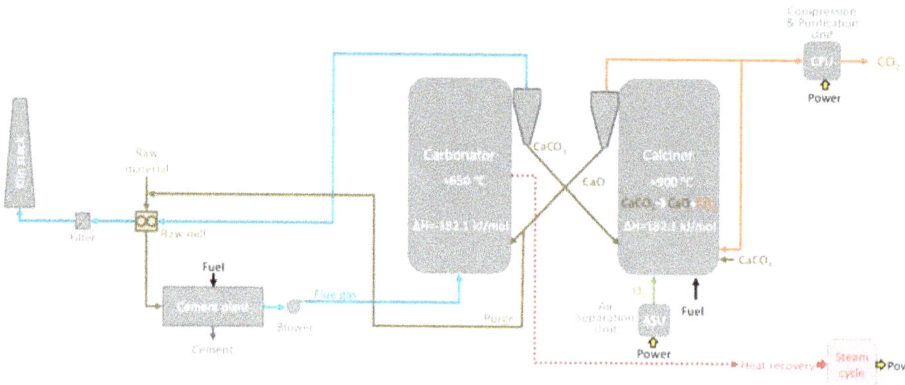

Figure 11. Simplified diagram flow of a CaL process installed in a cement plant: tail-end configuration (adapted from [46]).

In the carbonator, CaO is put into contact with the flue gas containing CO_2 at 600–700 °C. The $CaCO_3$ formed is then sent to the calcination reactor, where it is heated up to 890–930 °C, to recover the CO_2 product and to regenerate the CaO, which is sent back to the carbonator reactor [62]. In order to obtain a pure stream of CO_2, it is necessary to use oxyfuel combustion in the calciner, which has the drawback of the requirement of a cryogenic air separation unit (ASU), which significantly increases the CAPEX and OPEX. An emerging alternative is to heat the calciner indirectly, avoiding the need of an ASU. The outlet CO_2 enriched stream is sent to a compression and purification unit (CPU) where it can reach purities larger than 95% [63]. The sorbent tends to sinter and lose CO_2 capture capacity over cycles, so it is necessary to add a make-up stream of fresh $CaCO_3$ and to extract a purge stream rich in CaO to avoid the build-up of inert species [63]. However, the latter is not a problem if the CaL process is integrated in a cement plant that can make use of such purge as a feedstock for cement production (see Figure 11). This technology can offer capture rates up to 98% [64]. One of the inherent advantages of this technology is that a large part of the energy introduced into the calciner can be recovered as high temperature heat (≈650 °C) in the cooled carbonator to produce electricity with high efficiency [20,63].

There are two CaL configurations that could be implemented in the cement industry: the tail-end configuration (post-combustion technology), which has already been shown in Figure 11, and the integrated configuration, which is shown in Figure 12. In the integrated configuration, the calcination is integrated in the calciner of the cement kiln, which leads to a higher energy efficiency. Due to the small particle size required for the cement kiln, the integrated calciner is an entrained flow reactor [46]. Fuel consumption is larger in the tail-end option, although the high temperature heat can be recovered in a steam cycle to generate power. The specific energy consumption is larger for the tail-end technology: the SPECCA of a tail end configuration with 20% integration is 4.42 GJ/t CO_2 (4.07 GJ/t CO_2 for 50% integration), whereas for the fully integrated configuration the SPECCA is reduced to 3.17 GJ/t CO_2 [65]. The estimated cost related to the retrofitting of tail-end and integrated CaL to cement plants is 52.4 and 58.6 €/t CO_2 avoided, respectively [37].

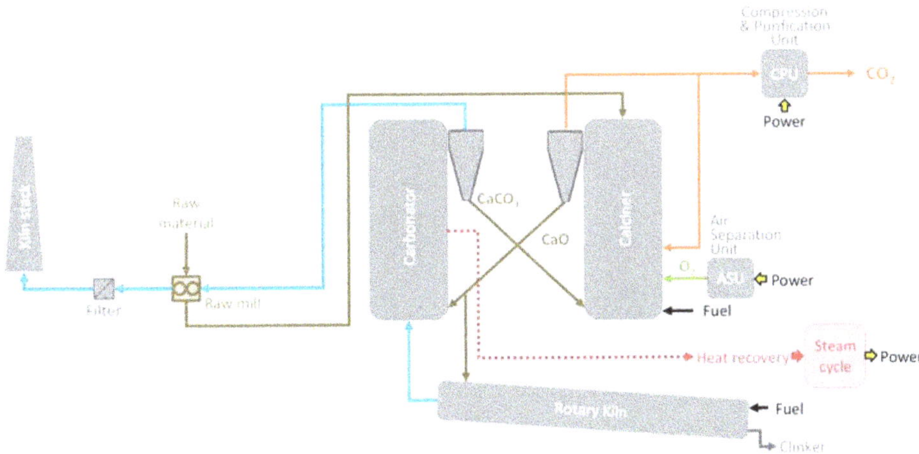

Figure 12. Simplified diagram flow of a CaL process installed in a cement plant: integrated configuration (adapted from [46]).

Industrial Technology Research Institute (ITRI) in cooperation with the Taiwan Cement Company (TCC) installed in 2013 a pilot CaL facility in the TCC's Ho Ping cement plant in Hualien, Taiwan, to demonstrate its High Efficiency Calcium Looping Technology (HECLOT). The pilot facility, which simplified flowsheet is shown in Figure 13, mainly consists on a bubbling fluidized bed carbonator with a diameter of 3.3 m and a height of 4.2 m, whose temperature was controlled by means of 36 water-cooled double steel jackets suspended from its top, and a rotary kiln calciner, with a diameter of 0.9 m and a length of 5 m. The pilot successfully captured 1 t CO_2/h from a slipstream of 3.1 t/h of flue gas with 20–25% CO_2 (equivalent to 1.9 MW_{th} out of 300 MW_{th} of the whole plant), with capture rates above 85% [66]. This is the world largest CaL pilot plant. In 2014, ITRI received an R&D 100 Award for this technology breakthrough in the environmental technologies category. According to ITRI, the estimated cost of the technology is less than \$30/t CO_2 [67]. The design of cascade-cyclone can increase the mixing efficiency of limestone and hot stream to increase the calcination efficiency above 90%. Oxyfuel combustion cyclone calciner with flue gas recirculation can reduce energy consumption and produce CO_2 purities above 90%. Integration of steam hydration can increase CO_2 capture efficiency and sorbent carbonation conversion, reducing sorbent consumption, recirculation rate, and thus the volume of the reactors and the energy consumption. The economic and feasible scale is considered to be a 50 MW demonstration plant [68]. TCC aims to reach zero emissions at its industrial facilities, and is developing HECLOT technology at the Ho Ping Cement plant not only to capture CO_2 but to reduce the cost of cement production. It has committed US \$19.1 M to expand the technology, and it estimates that by 2025 it will capture 0.45 Mt CO_2/y [69].

GE's RCC technology was evaluated within the Norcem CO_2 capture project (2013–2017). Pilot testing was conducted at the Institute of Combustion and Power Plant Technology at the University of Stuttgart (IFK), in Germany, under representative operation conditions focusing on the validation of process models [70]. The end of the pipe benchmarking carried out for Brevik, indicated a SPECCA of 3.13 GJ/t CO_2 avoided. Oxyfuel combustion RCC integration options provide attainable solutions for efficiently capturing CO_2 form cement plants, with lower specific energy consumption (SPECCA of 1.45 GJ/t CO_2 avoided). Emerging indirectly heated RCC concepts provide prospects to lower the overall cost by the ASU elimination [71].

Within CEMCAP project (2015–2018), the tail-end CaL (fluidized bed) was evaluated at two existing pilot facilities: the 200 kW_{th} rig at IFK, and the 30 kW_{th} pilot at INCAR-CSIC, in Spain [72]. CO_2 capture efficiencies up to 98% were attained. Testing involved the evaluation of different operation conditions, such as high CO_2 concentration (up to 33 vol %; wet basis), high limestone make-up ratios, looping ratios up to 20 mol CaO/mol CO_2, and carbonator temperatures between 600 and 710 °C. Tail-end CaL was demonstrated under industrially relevant conditions, and is considered to be ready for the demonstration at larger scale in the cement industry [73].

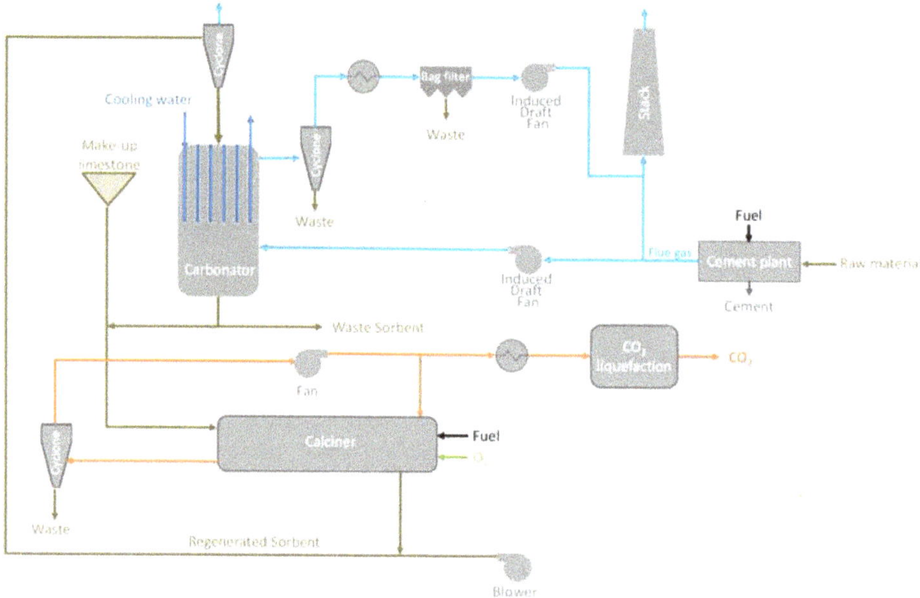

Figure 13. Set-up of the 1.9 MW_{th} HECLOT pilot facility installed at Ho Ping cement plant in Taiwan (adapted from [66]).

The integrated CaL configuration, or entrained flow CaL, is a promising option, although with a lesser degree of maturity than that of the tail-end CaL configuration [48]. This technology is being further developed within the European funded *CLEAN clinKER* production by calcium looping process (CLEANKER) project (2017–2021), which aims at demonstrating the entrained flow CaL technology at TRL 7 through the design, construction and operation of a pilot system in the Buzzi Unicem's cement plant in Vernasca, Italy. The pilot plant includes an entrained-flow carbonator with a 250 mm diameter in the up-flow section and 350 mm diameter in the downstream section, which treats a slipstream of approximately 1000 Nm^3/h of the kiln flue gas, and an entrained-flow oxyfuel calciner, connected to the kiln. The carbonator operates as an adiabatic reactor with an inlet adiabatic mixing temperature of 600 °C. The temperature rise due to the exothermic carbonation reaction is mitigated by heat losses and by air in-leakages. The pilot plant uses as sorbent the same raw meal that is used for cement production [74]. The project intends to demonstrate that CaL is an optimum alternative for CCUS in cement industry, with CO_2 capture efficiencies larger than 90%, ensuring that the clinker quality is not affected, and setting the basis for the industrial exploitation of the technology [75,76]. The two first years of the project were devoted to the detailed design and construction of the CaL demonstration system, and to the characterization of raw meals. Up to the project closure in September 2021, efforts are focused in the experimental campaigns, analysis of results, evaluation of the scale-up of the technology, economic analysis, and the life cycle assessment of the integrated system. The project also comprises mineralization tests with the captured CO_2 [74].

2.2. *State of the Art of Oxyfuel Combustion CO_2 Capture Technologies in the Cement Sector*

The configuration of an oxyfuel combustion cement plant is shown in Figure 14. Oxyfuel combustion refers to the combustion of the fuel in oxygen, instead of air. To control the temperature in the kiln, part of the flue gas must be recirculated (see Figure 14). As a consequence, both the material conversion in the kiln and the operational specifications of the overall process differ from those of conventional kilns [77]. The kiln exhaust would be mainly composed of CO_2 and steam, which can be easily knocked out by condensation. The resulting CO_2 stream (up to 80% mol CO_2, dry basis) would only need to be purified in a relatively simple cryogenic CPU unit to achieve 95% CO_2. Higher purities could be achieved by cryogenic distillation [4].

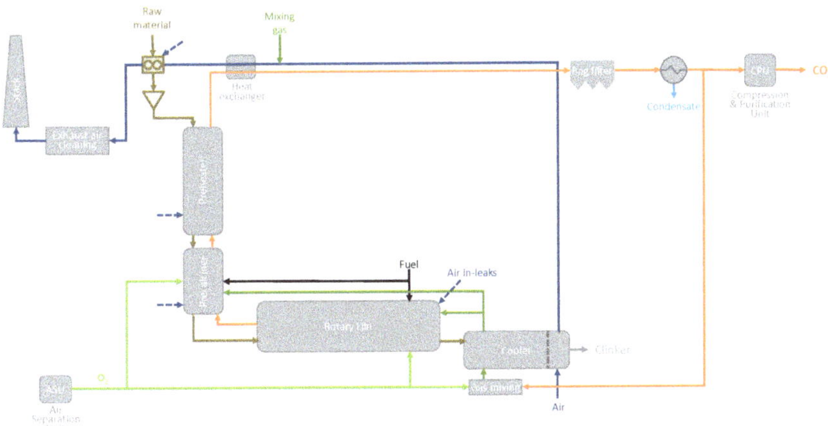

Figure 14. Configuration of an oxyfuel cement plant (adapted from [77]).

Oxyfuel combustion improves fuel efficiency, and provides a relatively low cost option for CO_2 abatement in cement plants compared to other technologies. Nevertheless, this method implies re-building and re-engineering the cement plant in order to minimize air ingress and optimize the heat recovery system. The main economic and energetic penalties arise from the need of an ASU, and the need to develop new kilns adapted to oxyfuel combustion conditions [78].

There is experience of oxygen enrichment in cement plants to improve throughput, or to enable the use of alternative fuels, but not for CO_2 abatement, and not at high levels of O_2 enrichment. Moreover, oxyfuel combustion capture has not been demonstrated yet at commercial scale in other sectors, and has an earlier stage of development than post-combustion capture technologies [4].

The European Cement Research Academy (ECRA) has been conducting research on oxyfuel combustion carbon capture since 2007. ECRA's CCS project phases I, II, and III have been completed, and phase IV is underway [79]. This might include the demonstration of oxyfuel technology at industrial scale in two European cement plants: Heilderberg Cement's plant in Colleferro, Italy, and Lafarge Holcim's plant in Retznei, Austria, which was announced in 2018 [80]. The project aims to advance the technology to TRL 7-8.

CEMCAP project (2015–2018), involved oxyfuel testing: a prototype that used CO_2 for cooling the clinker was manufactured by IKN and evaluated at a Heidelberg Cement's plant in Hannover; and an oxyfuel burner prototype, POLFLAME, was constructed by Thyssenkrupp based on a commercial kiln burner design. An existing oxyfuel combustion facility at IFK was also modified to entail testing under representative conditions of cement manufacturing. The results showed that the clinker quality and the cement strength were not affected by the oxyfuel operating conditions [48], advancing the technology to TRL 6. The estimated cost for the retrofit of oxyfuel combustion technology to cement plants is 42.4 €/t CO_2 avoided [37].

2.3. State of the Art of Direct CO_2 Capture Technologies in the Cement Sector

Low Emissions Intensity Lime and Cement (LEILAC), is a European funded project running between 2016 and 2020, which aims at demonstrating direct capture of process emissions at the lime and cement industries. A 60 m high pilot plant was built at Heidelberg Cement's plant in Lixhe, Belgium, with a capacity of 25,000 t CO_2/y (240 t/d of raw meal feed for cement production, or 190 t/d ground limestone), which is the equivalent to 5% of the factory's total CO_2 emissions. It makes use of Calix's technology, in which the conventional calciner is re-engineered to indirectly heat the limestone, using a direct separator reactor (DSR), which is a special steel tube that acts as a large heat exchanger (see Figure 15). In the pilot, natural gas is burned along the external furnace, heating the exterior of the DSR up to approximately 1000 °C. The raw material is dropped in the interior of the DSR, from its top, and falls slowly down, being heated by both conductive and radiative heat transfer from the reactor wall, causing it to calcine, releasing CO_2. At the base, the solids and gases are separated [81]. This way, the process emissions arising from the limestone calcination are isolated from those resulting from fuel combustion, which can be addressed using alternative fuels or with other CCUS technologies. The main advantage of the technology is its low energy penalty, which is only related with heat losses. Moreover, according to the developers of the technology, the capital cost is comparable to conventional equipment. On the other hand, its main challenge is the high temperature required for limestone calcination, which is significantly above than that of other applications of Calix's direct CO_2 separation technology. During pilot testing, special attention is being addressed to ensure that enough heat is put into the DSR and that the heat is placed at the right places. The capture rate is being evaluated from the difference in the CO_2 content of the powder before and after the DSR. Early results from the pilot plant are promising, although testing will be extended until the end of 2020 to gradually increase operational throughputs, temperatures and evaluate materials durability. LEILAC's has advanced the technology to a TRL of 6–7.

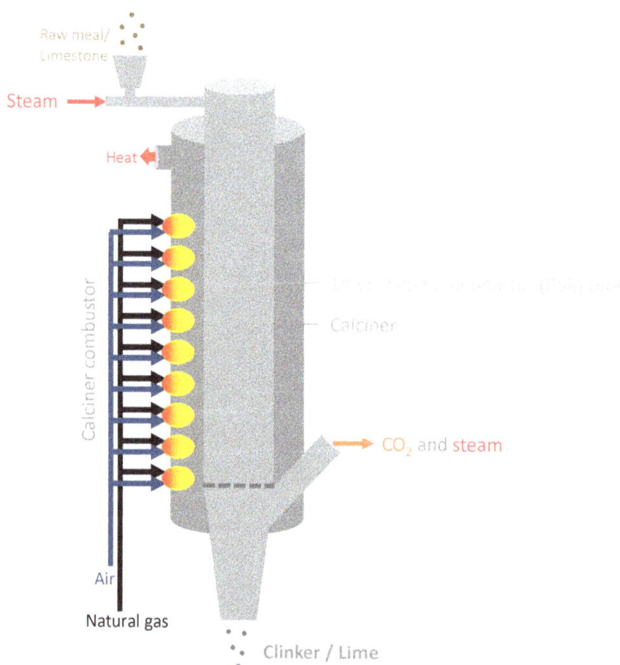

Figure 15. Scheme of LEILAC's direct separator reactor (adapted from [81]).

Based on the success of LEILAC project, LEILAC 2 was launched on 7 April 2020, and will run to the end of 2024. Central aspects of the LEILAC2 project are the further scale-up of the technology to an industrial level, the full process integration into an existing cement plant, and the heat supply by means of renewable energy to achieve carbon neutral CO_2 separation [14]. A CO_2 separation plant with capacity to capture 100,000 t CO_2/y (four-fold scale-up of LEILAC's pilot), which represents approximately 20% of a cement plant's process CO_2 emissions, or 100% of a large lime kiln's process emissions, will be built in a HeidelbergCement's plant in Western Europe, although the actual location is yet to be determined [82].

2.4. Comparison of CO_2 Capture Technologies Evaluated or Planned in the Cement Sector

Table 1 summarizes relevant information of the CO_2 capture technologies that have been evaluated in the cement sector, or those that are under planning stage. As can be seen from the table, the activity is concentrated in the last 7 years.

Caution is needed when comparing energy consumption and cost between technologies, due to the different assumptions of the technology developers. Costs depend strongly on factors such as the geographical location, the carbon tax, the steam source, the electricity mix, the electricity price, the fuel price, and the plant specific characteristics. To choose the best suited technology for a specific plant, specific technical and economic evaluations need to be performed. However, the efforts of certain projects, like Norcem CO_2 capture project and CEMCAP, which have compared different capture technologies based on a common basis, have provided relatively fair comparisons.

As shown in Table 1, within CEMCAP project, oxyfuel combustion CO_2 capture was found to provide the lowest cost per CO_2 avoided, followed by CaL, CAP, and MAL. Oxyfuel combustion CO_2 capture also showed the lowest specific primary energy consumption per CO_2 avoided (SPECCA), followed by integrated CaL, MAL, CAP, and tail-end CaL.

The lowest cost of CO_2 capture reported in Table 1 corresponds to HECLOT technology, with \$30/t CO_2. However, this value, which was taken from an ITRI's publication, has not been peer reviewed and must be taken with caution.

The second lowest cost reported in Table 1 corresponds to the CDRMax® technology. However, we must bear in mind that this cost is not site specific for the Dalmia cement's plant but a generic cost provided by Carbon Clean, and thus must also be taken with caution.

The third lowest cost shown in Table 1 corresponds to RTI solid sorbent technology, which was evaluated within Norcem CO_2 capture project.

The cost of Svante's technology is also amongst the lowest shown in Table 1. This has been taken from a public declaration of Svante's CEO regarding the commercial Lafarge Holcim cement CO_2 capture project, which has not been peer review, and thus must also be taken with caution.

However, cost is not the only factor to consider for the retrofit of a cement plant with CO_2 capture: technology maturity, possible effects over product quality, space requirements, and the need for utilities are important factors that need to be considered.

To date, chemical absorption with liquid solvents have reached the largest demonstration scale in the cement sector, with the SkyMine process at the front, with 75,000 t CO_2/y, followed by amine-based Anhui Conch's project, with 50,000 t CO_2/y. However, no operational performance data of those facilities are publicly available. On the other hand, amine absorption has reached commercial scale demonstration in the industry and power sector, and hence is considered the most mature technology. In the Longship Project, this has swayed the argument towards Aker Solution's ACC™ technology over RTI solid sorbent technology, despite the lower energy consumption of the later shown within Norcem CO_2 capture project (see Table 1).

Figure 16 summarizes the current TRL of CO_2 capture technologies in the cement sector. On one end we find membranes, with the lowest TRL, 4, and in the opposite end, chemical absorption, with TRL 8. In between, the trio formed by adsorption, oxyfuel, and direct capture, with TRL 6, and CaL, with TRL 7.

Table 1. Summary of carbon capture technologies evaluated and planned in the cement sector

Technology	Project	Test Location	Operation	Scale of Field Test	Type	Energy Consumption	Cost
SkyMine®	SkyMine® Beneficial CO_2 Use Project	Capitol Aggregates' San Antonio cement plant, Texas, United States of America	2015–	76,488 t CO_2/y	CCU	8.3 GJ_e/t CO_2 [23]	\$125 M [a] [23]
Aker Solution's ACC™	Norcem CO_2 capture project	Norcem's cement plant, Brevik, Norway	2014–2015	450 Nm^3/h kiln flue gas 140–250 kg CO_2/h	capture	3.0 GJ/t CO_2 [b,c,d] 2.7 GJ/t CO_2 [b,d] 0 GJ/t CO_2 [b] [36]	NA
Aker Solution's ACC™	Norway's full chain CCS	Norcem's cement plant, Brevik, Norway	2023–	400,000 t CO_2/y	CCS	NA	NA
Amine scrubbing	Anhui Conch Cement project	Anhui Conch's cement plant in Wuhu, China	2018–	50,000 t CO_2/y	CCU	NA	\$10 M
CDRMax®	Dalmia Cement project	Dalmia cement's plant in Tamil Nadu, India	NA	500,000 t CO_2/y	CCU	NA	\$40/t CO_2 [44]
CAP	CEMCAP	GE's Technology Center in Växjö, Sweden	2015–2018	-	Individual components tested: absorber, DCC-scrubber and water wash	2.37 GJ/t CO_2 [b] [47] 3.75 GJ/t CO_2 [e] [46]	66.2 €/t CO_2 [37]
Membranes	Norcem CO_2 capture project	Norcem's cement plant, Brevik, Norway	2014	NA	capture	NA	NA
Membranes	MemCCC project	Norcem's cement plant, Brevik, Norway	2016	37 Nm^3/h kiln flue gas	capture	1.20 GJ_e/t CO_2 [51]	47–48 €/t CO_2 [d] [51]
MAL	CEMCAP	TNO, Eindhoven, Netherlands and SINTEF Energy Research, Trondheim, Norway	2015–2018	-	Individual components tested separately: membranes and liquefaction	3.22 GJ_e/t CO_2 [e] [46]	83.5 €/t CO_2 [37]
RTI solid sorbent technology	Norcem CO_2 capture project	Norcem's cement plant, Brevik, Norway	2016	NA	capture	2.4 GJ/t CO_2 [b,c,d] [56]	45.8 €/t CO_2 [c,d] 40.7 €/t CO_2 [d] 38.6 €/t CO_2 [56]
Veloxotherm™	LafargeHolcim Cement Carbon Capture	Holcim Portland cement plant, Colorado, United States of America	2023–	2,000,000 t CO_2/y	CCUS (EOR)	NA	\$150 M \$50/t CO_2 [60]

Table 1. *Cont.*

Technology	Project	Test Location	Operation	Scale of Field Test	Type	Energy Consumption	Cost
Veloxotherm™	CO2MENT	Lafarge's Richmond British Columbia cement plant, Canada	2019–2022	1 t CO_2/d	CCU	NA	$28 M
HECLOT	ITRI's Calcium Looping Pilot	Ho Ping Cement plant, Taiwan	2013–	1 t CO_2/h	capture	NA	$30/t CO_2 [67]
RCC	Norcem CO_2 capture project	Institute of Combustion and Power Plant Technology at the University of Stuttgart (IFK), Germany	2013–2017	-	-	3.13 GJ/t CO_2 [e,h] 1.45 GJ/t CO_2 [e,i] [71]	NA
CaL	CEMCAP	IFK, Stuttgart University, Germany INCAR-CSIC, Spain	2015–2018	-	-	4.42 GJ/t CO_2 [e,f,h] 4.07 GJ/t CO_2 [e,g,h] 3.17 GJ/t CO_2 [e,i] [65]	52.4 €/t CO_2 [h] 58.6 €/t CO_2 [i] [37]
Entrained CaL	CLEANKER	Buzzi Unicem Vernasca cement plant, Italy	2019–2021	1000 Nm^3/h kiln flue gas	capture and mineralization testing	NA	NA
Oxyfuel combustion	CEMCAP	HeidelbergCement plant, Hannover	2015–2018	-	oxyfuel prototype testing: clinker cooler and burner	1.63 GJ/t CO_2 [e]	42.4 €/t CO_2 [37]
Oxyfuel combustion	ECRA's CCS project	HeidelbergCement plant in Coleferro, Italy Lafarge Holcim plant in Retznei, Austria	NA	NA	NA	NA	80 M€ [80]
Direct capture	LEILAC	Lixhe, Belgium	2020	25,000 t CO_2/y	NA	NA	NA
Direct capture	LEILAC 2	Heidelberg cement plant in Western Europe	2020–2024	100,000 t CO_2/y	NA	NA	NA

[a] Installed costs; [b] Specific reboiler duty (SRD); [c] Without heat recovery; [d] 85% capture rate; [e] Specific primary energy consumption per CO_2 avoided (SPECCA); [f] 20% integration; [g] 50% integration; [h] tail-end; [i] integrated; NA: Not available.

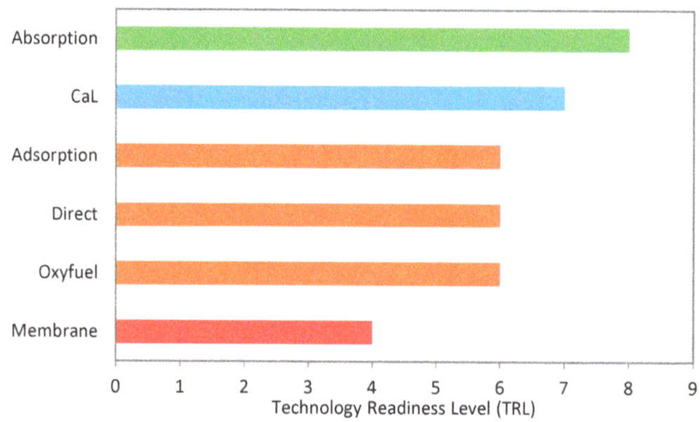

Figure 16. Current technology readiness level of CO_2 capture technologies in the cement sector (adapted from [83]).

3. Future Challenges and Prospects for CO_2 Capture in the Cement Industry

CO_2 capture is a necessity in the cement sector, where up to 65% of CO_2 emissions come from the manufacturing process, and thus cannot be avoided by other means. The Faster Innovation Case (FIC) recently published by the IEA, which is consistent with reaching net-zero emissions in 2050, would see 5 CCUS facilities of 1 Mt/y each month in the cement sector through 2050. Moreover: the increasing share of bioenergy in the fuel mix of cement kilns, which does not require significant equipment retrofits, if combined with CCUS, could lead to net CO_2 removals from the atmosphere, thus contributing to the achievement of net-zero emission across the entire economy [84].

CO_2 capture in the cement sector is now close to commercial demonstration. There are four large scale projects at different stages of development: Norway's Longship Project, which will make use of amine-based Aker solutions' ACC^{TM} technology to capture 0.4 Mt CO_2/y by 2023; Dalmia cement project, which will make use of another amine scrubbing technology, CDRMAX®, to capture 0.5 Mt CO_2/y; LafargeHolcim cement carbon capture, which would make use of the Svante's adsorption-based Veloxotherm^{TM} process to capture up to 2 Mt CO_2/y; and TCC is developing HECLOT technology to capture 0.45 M tCO_2 by 2025. Although at a lower scale, LEILAC 2 will also demonstrate direct CO_2 capture at the significant scale of 0.1 Mt CO_2/y by 2024.

While direct CO_2 capture and oxycalcination are promising technologies that have great potential to be implemented in new cement plants, the prospects for their use in the retrofitting of existing plants is less likely. Given the existing overcapacity, and the age and life ahead of existing facilities, the construction of new cement plants integrated with CCUS at a relevant scale for climate mitigation seems unlikely.

Post-combustion CO_2 capture technologies are the preferred option to retrofit existing facilities, as the CO_2 is captured from the exhaust gas of the cement plant, thus not affecting the existing cement production process. Among the available post-combustion technologies, chemical absorption with liquid solvents is the most mature technology: to date, it has reached the largest demonstration scale at cement plants, and provide the least risky pathway for the retrofitting of existing facilities. Moreover, solvent and process development are expected to lead to further cost reductions as the technology deploys in the cement sector, as it has previously occurred in the power sector. In fact, this is the technology selected by two of the commercial-scale projects in the pipeline. On the other hand, although with a lesser technology readiness level, solid sorbents based post-combustion CO_2 capture processes also show great promise. Svante's adsorption and HECLOT technologies are already on commercial scale development.

Further research and development is urgently needed in order to reduce the cost of the capture step and to increase the technology readiness level of emerging technologies to make CCUS an economically viable and safer option for cement producers in the forthcoming CO_2 neutral economy.

Author Contributions: Conceptualization, M.G.P.; Writing—original draft preparation, S.M.; Writing—review and editing, M.G.P. and F.R.; Visualization, M.G.P.; Supervision, M.G.P. All authors have read and agreed to the published version of the manuscript.

Funding: S.M. acknowledges the award of an Introduction to Research fellowship (grant number: JAEICU-19-INCAR-26) from the JAE Intro ICUs Programme of the Agencia Estatal Consejo Superior de Investigaciones Científicas (CSIC). M.G.P. acknowledges support from the Ramon y Cajal Programme (grant number: RyC-2015-17516) of the Government of Spain, co-financed by the European Social Fund. Financial support from the Government of the Principado de Asturias (PCTI, grant number: IDI/2018/000115), co-financed by the European Regional Development Fund (ERDF), is also acknowledged.

Conflicts of Interest: The authors declare no conflict of interest.

Abbreviations

2DS	IEA's 2 °C scenario
ACC™	Aker Solution's Advanced Carbon Capture technology
AIPOs	Alumino-phosphates
ASU	Air separation unit
ASTR	Automated sorbent test rig
B2DS	IEA's Beyond 2 °C scenario
CaL	Calcium looping
CAP	Chilled Ammonia Process
CAPEX	Capital expenditure
CCP	Carbon Capture Project
CCS	CO_2 capture and storage
CCU	CO_2 capture and utilization
CCUS	CO_2 capture, utilization and storage
CHP	Combined heat and power facility
CLEANKER	Clean clinker production by calcium looping process
CPU	Compression and purification unit
CSA	Concentration swing adsorption
DCC	Direct contact cooler
DOE-NETL	United States Department of Energy's National Energy Technology Laboratory
CDRMax®	Carbon Clean's amine-based CO_2 capture technology
CEMBUREAU	The European Cement Association
DEA	Diethanolamine
DSR	Direct separator reactor
ECRA	European Cement Research Academy
EHR	Excess heat recovery
EOR	Enhanced oil recovery
ESP	Electrostatic precipitator
FGD	Flue gas desulfurization
FSC	Fixed-site carrier
GCCA	Global Cement and Concrete Association
IEA	International Energy Agency
IFK	Institute of Combustion and Power Plant Technology at the University of Stuttgart
ITRI	Industrial Technology Research Institute
HECLOT	High Efficiency Calcium Looping Technology
LEILAC	Low Emissions Intensity Lime and Cement
MAL	Membrane-assisted liquefaction
MEA	Monoethanolamine
MOFs	Metal organic frameworks
MTU	Mobile test unit
NDC	National determined contributions
NTNU	Norwegian University of Science and Technology

OPEX	Operational expenditures
PSA	Pressure swing adsorption
PVAm	Polyvinylamine
RCC	GE's Regenerative Calcium Cycle
RTI	Research Triangle Institute
RTS	IEA's Reference Technology Scenario
SAPOs	Alumino-silico-phosphates
SFMB	Staged fluidized moving bed
SNCR	Selective non catalytic reduction
SPECCA	Specific primary energy consumption for CO_2 avoided
TCC	Taiwan Cement Company
TRL	Technology readiness level
TSA	Temperature swing adsorption
VSA	Vacuum swing adsorption
VTCSA	Vacuum, temperature, and concentration swing adsorption

Symbols and Colors Notation

References

1. Edwards, P. The 2010's: A decade in the cement sector. *Global Cement*, 5 December 2019; pp. 10–17.
2. IEA. Cement. Available online: https://www.iea.org/reports/cement (accessed on 30 July 2020).
3. IEA; CSI. *Technology Roadmap Low-Carbon Transition in the Cement Industry*; IEA: Paris, France, 2018.
4. IEAGHG. *CO_2 Capture in the Cement Industry*; IEAGHG: Cheltenham, UK, 2008.
5. Bellona. *Climate Action in the Cement Industry*; Bellona: Oslo, Norway, 2020.
6. IEA. *World Energy Outlook 2019*; International Energy Agency (IEA): Paris, France, 2019; ISBN 978-92-64-97300-8.
7. Andrew, R.M. Global CO_2 emissions from cement production. *Earth Syst. Sci. Data* **2018**, *10*, 195–217. [CrossRef]
8. Hepburn, C.; Adlen, E.; Beddington, J.; Carter, E.A.; Fuss, S.; Mac Dowell, N.; Minx, J.C.; Smith, P.; Williams, C.K. The technological and economic prospects for CO_2 utilization and removal. *Nature* **2019**, *575*, 87–97. [CrossRef] [PubMed]
9. IPCC. *Global Warming of 1.5 °C: An IPCC Special Report on the Impacts of Global Warming of 1.5 °C above Pre-Industrial Levels and Related Global Greenhouse Gas Emission Pathways, in the Context of Strengthening the Global Response to the Threat of Climate Change, Sustainable Development, and Efforts to Eradicate Poverty*; Masson-Delmotte, V., Zhai, P., Pörtner, H.-O., Roberts, D., Skea, J., Shukla, P.R., Pirani, A., Moufouma-Okia, W., Péan, C., Pidcock, R., et al., Eds.; IPCC: Geneva, Switzerland, 2018.
10. GCCA. GCCA Launches 'Innovandi—The Global Cement and Concrete Research Network'. Singapore, 10 October 2019. Available online: https://gccassociation.org/news/gcca-launches-innovandi-the-global-cement-and-concrete-research-network/ (accessed on 1 August 2020).
11. GCCA. Innovandi—The Global Cement and Concrete Research Network. Available online: https://gccassociation.org/innovandi/ (accessed on 1 August 2020).
12. CEMBUREAU. *Cementing the European Green Deal*; The European Cement Association: Brussels, Belgium, 2020.
13. Beumelburg, C. HeidelbergCement First Cement Company to Receive Approval for Science-Based CO_2 Reduction Targets. 13 May 2019. Available online: https://www.heidelbergcement.com/en/pr-13-05-2019 (accessed on 6 May 2020).
14. HeidelbergCement. HeidelbergCement and Partners Drive Innovative CO_2 Separation. 30 March 2020. Available online: https://www.heidelbergcement.com/en/pr-30-03-2020 (accessed on 6 August 2020).
15. GCCSI. *The Global Status of CCS: 2019*; Global CCS Institute: Melbourne, Australia, 2019.
16. IPCC. *IPCC Special Report on Carbon Dioxide Capture and Storage*; Metz, B., Davidson, O., Coninck, H.C.D., Loos, M., Meyer, L.A., Eds.; Cambridge University Press: Cambridge, UK; New York, NY, USA, 2005; ISBN 13 978-0-521-86643-9.
17. IEA. *Transforming Industry through CCUS*; International Energy Agency: Paris, France, 2019.
18. Naranjo, M.; Brownlow, D.T.; Garza, A. CO_2 capture and sequestration in the cement industry. *Energy Procedia* **2011**, *4*, 2716–2723. [CrossRef]
19. ECRA. *Carbon Capture Technology—Options and Potentials for the Cement Industry*; TR 044/2007; European Cement Research Academy GmbH: Düsseldorf, Germany, 2007.
20. Hills, T.; Leeson, D.; Florin, N.; Fennell, P. Carbon Capture in the Cement Industry: Technologies, Progress, and Retrofitting. *Environ. Sci. Technol.* **2015**, *50*. [CrossRef] [PubMed]
21. CEMBUREAU. *The Role of Cement in the 2050 Low Carbon Economy*; The European Cement Association: Brussels, Belgium, 2018.
22. Bjerge, L.-M.; Brevik, P. CO_2 Capture in the Cement Industry, Norcem CO_2 Capture Project (Norway). *Energy Procedia* **2014**, *63*, 6455–6463. [CrossRef]
23. Walters, J. *SkyMine® Beneficial CO_2 Use Project. Final Report*; Skyonic Corporation: Austin, TX, USA, 2016.
24. Jones, J.; Barton, C.; Clayton, M.; Yablonsky, A.; Legere, D. *SkyMine® Carbon Mineralization Pilot Project. Final Phase 1 Topical Report*; Skyonic Corporation: Austin, TX, USA, 2011.
25. Bottoms, R.R. Process for Separating Acidic Gases. U.S. Patent 1783901, 2 December 1930.
26. IEAGHG. *Integrated Carbon Capture and Storage Project at Saskpower's Boundary Dam Power Station*; IEA Environmental Projects Ltd.: Cheltenham, UK, 2015.

27. Miyamoto, O.; Maas, C.; Tsujiuchi, T.; Inui, M.; Hirata, T.; Tanaka, H.; Yonekawa, T.; Kamijo, T. KM CDR Process™ project update and the new novel solvent development. *Energy Procedia* **2017**, *114*, 5616–5623. [CrossRef]
28. Dubois, L.; Thomas, D. Comparison of various configurations of the absorption-regeneration process using different solvents for the post-combustion CO_2 capture applied to cement plant flue gases. *Int. J. Greenh. Gas Control* **2018**, *69*, 20–35. [CrossRef]
29. Wang, M.; Lawal, A.; Stephenson, P.; Sidders, J.; Ramshaw, C. Post-combustion CO_2 capture with chemical absorption: A state-of-the-art review. *Chem. Eng. Res. Des.* **2011**, *89*, 1609–1624. [CrossRef]
30. Kim, S.; Shi, H.; Lee, J.Y. CO_2 absorption mechanism in amine solvents and enhancement of CO_2 capture capability in blended amine solvent. *Int. J. Greenh. Gas Control* **2016**, *45*, 181–188. [CrossRef]
31. Dutcher, B.; Fan, M.; Russell, A.G. Amine-based CO_2 capture technology development from the beginning of 2013—A review. *ACS Appl. Mater. Interfaces* **2015**, *7*, 2137–2148. [CrossRef]
32. Knudsen, J.N.; Bade, O.M.; Askestad, I.; Gorset, O.; Mejdell, T. Pilot plant demonstration of CO_2 capture from cement plant with advanced amine technology. *Energy Procedia* **2014**, *63*, 6464–6475. [CrossRef]
33. Bartolomé Muñoz, C.; Mora Peris, P.; Recalde Rodríguez, J.D. *Estado del Arte de las Tecnologías de Captura y Almacenamiento de CO_2 en la Industria del Cemento*; Agrupación de Fabricantes de Cemento de España: Madrid, Spain, 2011; ISBN 978-84-615-5702-8.
34. Vega, F.; Baena-Moreno, F.M.; Gallego Fernández, L.M.; Portillo, E.; Navarrete, B.; Zhang, Z. Current status of CO_2 chemical absorption research applied to CCS: Towards full deployment at industrial scale. *Appl. Energy* **2020**, *260*, 114313. [CrossRef]
35. IEAGHG. *Deployment of CCS in the Cement Industry*; IEA Environmental Projects Ltd.: Cheltenham, UK, 2013.
36. Knudsen, J.N. Results and Future Perspective of Aker Solutions' Amine Project. In Proceedings of the Norcem CO_2 Capture Project, Langesund, Norway, 20 May 2015.
37. Gardarsdottir, S.O.; De Lena, E.; Romano, M.; Roussanaly, S.; Voldsund, M.; Pérez-Calvo, J.-F.; Berstad, D.; Fu, C.; Anantharaman, R.; Sutter, D.; et al. Comparison of Technologies for CO_2 Capture from Cement Production—Part 2: Cost Analysis. *Energies* **2019**, *12*, 542. [CrossRef]
38. Government.no. The Government Launches 'Longship' for Carbon Capture and Storage in Norway. Norway, 21 September 2020. Available online: https://www.regjeringen.no/en/aktuelt/the-government-launches-longship-for-carbon-capture-and-storage-in-norway/id2765288/ (accessed on 13 October 2020).
39. Knudsen, J.C. Carbon Capture Utilization and Storage (CCUS). In Proceedings of the Carnegie ESG Seminar, Oslo, Norway, 21 November 2019.
40. AkerSolutions. Aker Solutions' Carbon Capture and Storage Technology Gets DNV GL Approval. Norway, 29 April 2020. Available online: https://www.akersolutions.com/news/news-archive/2020/aker-solutions-carbon-capture-and-storage-technology-gets-dnv-gl-approval/ (accessed on 30 July 2020).
41. GCCSI. World's Largest Capture Pilot Plant for Cement Commissioned in China. Available online: https://www.globalccsinstitute.com/news-media/insights/worlds-largest-capture-pilot-plant-for-cement-commissioned-in-china/ (accessed on 5 August 2020).
42. Conch. Conch Group's First Batch of Industrial Grade Carbon Dioxide Products Were Successfully Dispatched for Sale. 31 October 2018. Available online: http://www.conch.cn/en/News/info.aspx?itemid=166511 (accessed on 4 August 2020).
43. GCCSI. Dalmia Cement (Bharat) Limited and Carbon Clean Solutions Team up to Build Cement Industry's Largest Carbon Capture Plant. Available online: https://www.globalccsinstitute.com/news-media/latest-news/dalmia-cement-bharat-limited-and-carbon-clean-solutions-team-up-to-build-cement-industrys-largest-carbon-capture-plant/ (accessed on 24 July 2020).
44. CarbonClean. Solvents. Available online: https://www.carbonclean.com/solvents (accessed on 13 October 2020).
45. Lombardo, G.; Agarwal, R.; Askander, J. Chilled Ammonia Process at Technology Center Mongstad—First results. *Energy Procedia* **2014**, *51*, 31–39. [CrossRef]
46. Voldsund, M.; Gardarsdottir, S.O.; De Lena, E.; Pérez-Calvo, J.-F.; Jamali, A.; Berstad, D.; Fu, C.; Romano, M.; Roussanaly, S.; Anantharaman, R.; et al. Comparison of technologies for CO_2 capture from cement production—Part 1: Technical evaluation. *Energies* **2019**, *12*, 559. [CrossRef]
47. Augustsson, O.; Oskarsson, A.; Grubbström, J.; Sutter, D. *D10.4. Feasibility Study for CAP Process Scale-up*; GE Power Sweden: Norrköping, Sweden, 2018.

48. Jordal, K.; Abanades, C.; Cinti, G.; Berstad, D.; Hoenig, V.; Hornberger, M.; Monteiro, J.G.; Gardarsdottir, S.; Ruppert, J.; Størset, S.; et al. *D2.11. CEMCAP Strategic Conclusions—Progressing CO_2 Capture from Cement towards Demonstration Revision 1*; SINTEF Energi AS: Trondheim, Norway, 2019.
49. Baker, R.W.; Freeman, B.; Kniep, J.; Huang, Y.I.; Merkel, T.C. CO_2 capture from cement plants and steel mills using membranes. *Ind. Eng. Chem. Res.* **2018**, *57*, 15963–15970. [CrossRef]
50. He, X. A review of material development in the field of carbon capture and the application of membrane-based processes in power plants and energy-intensive industries. *Energy Sustain. Soc.* **2018**, *8*, 34. [CrossRef]
51. Hägg, M.B.; Lindbråthen, A.; He, X.; Nodeland, S.G.; Cantero, T. Pilot demonstration-reporting on CO_2 capture from a cement plant using hollow fiber process. *Energy Procedia* **2017**, *114*, 6150–6165. [CrossRef]
52. He, X. The latest development on membrane materials and processes for post-combustion CO_2 capture: A review. *SF J. Mater. Chem. Eng.* **2018**, *1*, 1009:1–1009:8.
53. SINTEF. CEMCAP. Available online: https://www.sintef.no/projectweb/cemcap/research/sp4/wp11-membrane-assisted-co2-liquefaction/ (accessed on 30 July 2020).
54. Sayari, A.; Belmabkhout, Y.; Serna-Guerrero, R. Flue gas treatment via CO_2 adsorption. *Chem. Eng. J.* **2011**, *171*, 760–774. [CrossRef]
55. Ruthven, D.M. *Principles of Adsorption and Adsorption Processes*; John Wiley and Sons: New York, NY, USA, 1984; ISBN 1-56081-517-5.
56. Nelson, T.O.; Kataria, A.; Mobley, P.; Soukri, M.; Tanthana, J. RTI's solid sorbent-based CO_2 capture process: Technical and economic lessons learned for application in coal-fired, NGCC, and cement plants. *Energy Procedia* **2017**, *114*, 2506–2524. [CrossRef]
57. Total. Svante, LafargeHolcim, Oxy Low Carbon Ventures and Total Launch Study for Commercial-Scale Carbon Capture and End-Use at U.S. Plant. Vancouver/Zurich/Houston/Paris, 6 January 2020. Available online: https://www.total.com/media/news/press-releases/svante-lafargeholcim-oxy-low-carbon-ventures-and-total-launch-study-commercial-scale-carbon-capture (accessed on 24 July 2020).
58. Total. U.S. Department of Energy's National Energy Technology Laboratory Announces Investment to Further Develop LH CO2MENT Colorado Project, Carbon Capture Technology. 17 September 2020. Available online: https://www.total.com/media/news/actualites/us-department-of-energys-national-energy-technology-laboratory-announces (accessed on 13 October 2020).
59. Svante. Available online: https://svanteinc.com/carbon-capture-technology/ (accessed on 29 June 2020).
60. Moore, A. *US 45Q Tax Credit Key to Developing Carbon Capture Facility in Colorado*; S&P Global Platts; 2020; Available online: https://www.spglobal.com/platts/en/market-insights/latest-news/coal/010820-us-45q-tax-credit-key-to-developing-carbon-capture-facility-in-colorado. (accessed on 13 October 2020).
61. LafargeHolcim. LafargeHolcim Launches Carbon Capture Project in Canada. Zug. Available online: https://www.lafargeholcim.com/lafargeholcim-launch-carbon-capture-project-canada (accessed on 24 July 2020).
62. Hornberger, M.; Spörl, R.; Scheffknecht, G. Calcium looping for CO_2 capture in cement plants—Pilot scale test. *Energy Procedia* **2017**, *114*, 6171–6174. [CrossRef]
63. de Lena, E.; Spinelli, M.; Romano, M.C. CO_2 capture in cement plants by "Tail-end" Calcium Looping process. *Energy Procedia* **2018**, *148*, 186–193. [CrossRef]
64. Cinti, G.; Matai, R.; Becker, S.; Alonso, M.; Abanades, C.; Spinelli, M.; De Lena, E.; Consonni, S.; Romano, M.; Hornberger, M.; et al. Options for Calcium Looping for CO_2 Capture in the Cement Industry. In Proceedings of the 2nd ECRA/CEMCAP Workshop: Carbon Capture Technologies in the Cement Industry, Dusseldorf, Germany, 6–7 November 2017.
65. de Lena, E.; Spinelli, M.; Gatti, M.; Scaccabarozzi, R.; Campanari, S.; Consonni, S.; Cinti, G.; Romano, M.C. Techno-economic analysis of calcium looping processes for low CO_2 emission cement plants. *Int. J. Greenh. Gas Control* **2019**, *82*, 244–260. [CrossRef]
66. Chang, M.-H.; Chen, W.-C.; Huang, C.-M.; Liu, W.-H.; Chou, Y.-C.; Chang, W.-C.; Chen, W.; Cheng, J.-Y.; Huang, K.-E.; Hsu, H.-W. Design and experimental testing of a 1.9 MW_{th} calcium looping pilot plant. *Energy Procedia* **2014**, *63*, 2100–2108. [CrossRef]
67. Hwang, Y. HECLOT, an Innovation for Circular Economy. *ITRI Today Spring*. 2020. Available online: https://itritoday.itri.org/100/content/en/unit_02-3.html (accessed on 4 August 2020).

68. ITRI. Calcium-Looping CO_2 Capture Technology. Available online: https://www.itri.org.tw/english/ListStyle.aspx?DisplayStyle=01_content&SiteID=1&MmmID=1037333532432522160&MGID=621024013054352667 (accessed on 4 August 2020).
69. CN. Taiwan Cement Co Sets CO_2 Capture Targets. *CemNet*. 12 June 2019. Available online: https://www.cemnet.com/News/story/166756/taiwan-cement-co-sets-co2-capture-targets.html (accessed on 4 August 2020).
70. Balfe, M.C.; Augustsson, O.; Tahoces-soto, R.; Bjerge, L.-M.H. Alstom's Regenerative Calcium Cycle—Norcem derisking study: Risk mitigation in the development of a 2nd generation CCS technology. *Energy Procedia* **2014**, *63*, 6440–6454. [CrossRef]
71. Balfe, M. Alstom's Regenerative Calcium Cycle. Results and Future Perspective. In Proceedings of the Norcem CO_2 Capture Project—Int. CCS Conference, Langesund, Norway, 20–21 May 2015.
72. Arias, B.; Alonso, M.; Abanades, C. CO_2 capture by calcium looping at relevant conditions for cement plants: Experimental testing in a 30 kW_{th} pilot plant. *Ind. Eng. Chem. Res.* **2017**, *56*, 2634–2640. [CrossRef]
73. Jordal, K.; Voldsund, M.; Størset, S.; Fleiger, K.; Ruppert, J.; Spörl, R.; Hornberger, M.; Cinti, G. CEMCAP—Making CO_2 capture retrofittable to cement plants. *Energy Procedia* **2017**, *114*, 6175–6180. [CrossRef]
74. Fantini, M. Clean and Green. *World Cement*. September 2019. Available online: http://www.cleanker.eu/clean-and-green-world-cement (accessed on 4 August 2020).
75. EC. CLEANKER CLEAN clinKER Production by Calcium Looping Process. Available online: https://ec.europa.eu/inea/en/horizon-2020/projects/h2020-energy/carbon-capture-storage-power-plants/cleanker (accessed on 24 July 2020).
76. CLEANKER. Clean Clinker by Calcium Looping for Low CO_2 Cement. Available online: http://www.cleanker.eu/ (accessed on 24 July 2020).
77. ECRA. *ECRA CCS Project: Report on Phase IV.A*; TR-ECRA-128/2016; European Cement Research Academy: Düsseldorf, Germany, 2016.
78. Ditaranto, M.; Bakken, J. Study of a full scale oxy-fuel cement rotary kiln. *Int. J. Greenh. Gas Control* **2019**, *83*, 166–175. [CrossRef]
79. ECRA. CCS—Carbon Capture and Storage. Available online: https://ecra-online.org/research/ccs/ (accessed on 5 August 2020).
80. ECRA. Cement Industry Launches an Industrial-Scale Carbon Capture Project. 29 January 2018. Available online: https://ecra-online.org/press-releases/ (accessed on 10 July 2020).
81. Hills, T.P.; Sceats, M.; Rennie, D.; Fennell, P. LEILAC: Low Cost CO_2 Capture for the Cement and Lime Industries. *Energy Procedia* **2017**, *114* (Suppl. C), 6166–6170. [CrossRef]
82. CCJ. LEILAC Project Moves to Phase 2. *Carbon Capture J.* **2020**, *75*, 8.
83. IEA. ETP Clean Energy Technology Guide. Available online: https://www.iea.org/articles/etp-clean-energy-technology-guide (accessed on 13 October 2020).
84. IEA. *Energy Technologies Perspectives 2020*; International Energy Agency: Paris, France, 2020.

energies

MDPI

Review

Recent Advances in Small-Scale Carbon Capture Systems for Micro-Combined Heat and Power Applications

Wahiba Yaïci [1,*], Evgueniy Entchev [1] and Michela Longo [2]

[1] CanmetENERGY Research Centre, Natural Resources Canada, 1 Haanel Drive, Ottawa, ON K1A 1M1, Canada; evgueniy.entchev@nrcan-rncan.gc.ca
[2] Department of Energy, Politecnico di Milano, Via La Masa, 34-20156 Milan, Italy; michela.longo@polimi.it
* Correspondence: wahiba.yaici@nrcan-rncan.gc.ca; Tel.: +1-613-996-3734

Abstract: To restrict global warming and relieve climate change, the world economy requires to decarbonize and reduce carbon dioxide (CO_2) emissions to net-zero by mid-century. Carbon capture and storage (CCS), and carbon capture and utilization (CCU), by which CO_2 emissions are captured from sources such as fossil power generation and combustion processes, and further either reused or stored, are recognized worldwide as key technologies for global warming mitigation. This paper provides a review of the latest published literature on small-scale carbon capture (CC) systems as applied in micro combined heat and power cogeneration systems for use in buildings. Previous studies have investigated a variety of small- or micro-scale combined heat and power configurations defined by their prime mover for CC integration. These include the micro gas turbine, the hybrid micro gas turbine and solid-state fuel cell system, and the biomass-fired organic Rankine cycle, all of which have been coupled with a post-combustion, amine-based absorption plant. After these configurations are defined, their performance is discussed. Considerations for optimizing the overall system parameters are identified using the same sources. The paper considers optimization of modifications to the micro gas turbine cycles with exhaust gas recirculation, humidification, and more advanced energy integration for optimal use of waste heat. Related investigations are based largely on numerical studies, with some preliminary experimental work undertaken on the Turbec T100 micro gas turbine. A brief survey is presented of some additional topics, including storage and utilization options, commercially available CC technologies, and direct atmospheric capture. Based on the available literature, it was found that carbon capture for small-scale systems introduces a large energy penalty due to the low concentration of CO_2 in exhaust gases. Further development is required to decrease the energy loss from CC for economic feasibility on a small scale. For the micro gas turbine, exhaust gas recirculation, selective gas recirculation, and humidification were shown to improve overall system economic performance and efficiency. However, the highest global efficiencies were achieved by leveraging turbine exhaust waste heat to reduce the thermal energy requirement for solvent regeneration in the CC plant during low- or zero-heating loads. It was shown that although humidification cycles improved micro gas turbine cycle efficiencies, this may not be the best option to improve global efficiency if turbine waste heat is properly leveraged based on heating demands. The biomass-organic Rankine cycle and hybrid micro gas turbine, and solid-state fuel cell systems with CC, are in early developmental stages and require more research to assess their feasibility. However, the hybrid micro gas turbine and solid-state fuel cell energy system with CC was shown numerically to reach high global efficiency (51.4% LHV). It was also shown that the biomass-fired organic Rankine cycle system could result in negative emissions when coupled with a CC plant. In terms of costs, it was found that utilization through enhanced oil recovery was a promising strategy to offset the cost of carbon capture. Direct atmospheric capture was determined to be less economically feasible than capture from concentrated point sources; however, it has the benefit of negative carbon emissions.

Keywords: carbon capture (CC); carbon capture and storage (CCS); micro-combined heat and power (micro-CHP); micro-cogeneration; energy systems; buildings; GHG emissions

Citation: Yaïci, W.; Entchev, E.; Longo, M. Recent Advances in Small-Scale Carbon Capture Systems for Micro-Combined Heat and Power Applications. *Energies* **2022**, *15*, 2938. https://doi.org/10.3390/en15082938

Academic Editors: Marta González Plaza and Rui P. P. L. Ribeiro

Received: 25 February 2022
Accepted: 13 April 2022
Published: 16 April 2022

1. Introduction

The International Energy Agency (IEA) has set a goal of carbon neutrality by 2100 for climate change mitigation. Additionally, the Paris Agreement has set a global initiative to reduce global warming to much less than 2 °C and to further engage efforts to reduce it to 1.5 °C [1–3]. Despite the increasing use of renewable energy sources, their growth is still insufficient to switch from fossil fuel consumption by the end of the century. Further, the current trend will limit the rise in global temperature to 2.7 °C by 2100, which is not sufficient to meet the 2 °C reduction target. Carbon capture, utilization, and storage (CCUS) is a proposed solution to achieve carbon neutrality during the transition from fossil fuels to fully renewable energy generation. Carbon capture (CC) is an attractive option as it aims to achieve carbon neutrality, while simultaneously generating waste carbon dioxide (CO_2) that can be converted into products and sold for profit. Further, decentralized energy generation has been widely investigated as a possible developmental path for achieving carbon neutrality. Distributed power generation in the form of small, decentralized systems can support decrease in emissions and protection of grid capacity, while also offering options for renewable energy [4–8].

The available options for CC are grouped into either post-combustion, pre-combustion, or oxy-fuel combustion [9–15]. The basic processes are summarized in Figure 1.

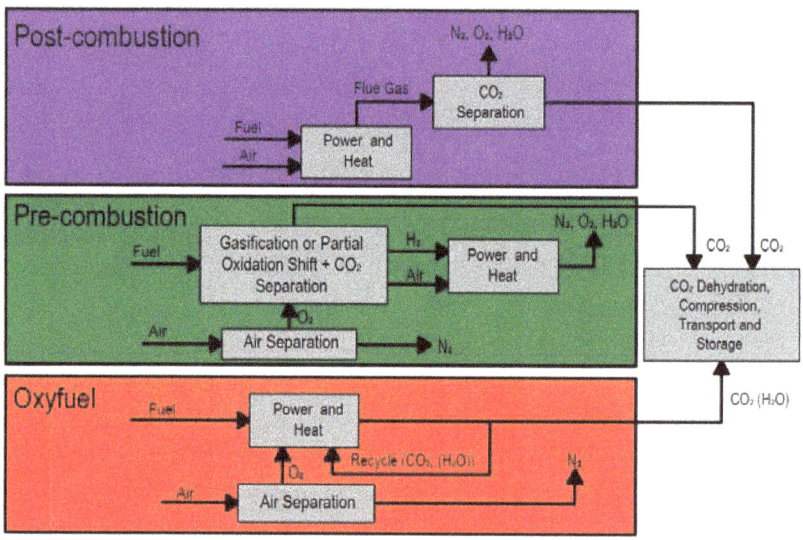

Figure 1. Overview of carbon capture options (adapted from [15]).

The capture methods that fall within each classification, as well as storage and utilization options, are summarized in Figure 2. Among the methods shown, absorption by mono-ethanolamine (MEA) for post-combustion capture is the most developed. Because of the high thermal energy requirement for solvent regeneration, this process is well suited to combined heat and power (CHP) applications, which supply recoverable heat [10]. Post-combustion absorption plants with an MEA-based solvent is the more commonly utilized method for small-scale carbon capture [16].

There exist numerous reviews and studies on large-scale carbon dioxide capture, utilization and storage [16–23], carbon capture and storage (CCS) [24–27], demonstration and deployment of CCS systems [28–33], cost of CCS [34,35], CCS applied in industry [36–38], CC [39,40], integration of CC in power generation plants [41,42] and in community scale energy systems [43,44], CC in pre-combustion, post-combustion and oxy-combustion in thermal power plants [45], CC post-combustion by chemical-looping [46], chemical absorption [46–51] or physical adsorption [52–54], membrane-based CC [55–57], and, finally, CC and separation technologies for end-of-pipe applications [58]. Although there is a myriad of dedicated research studies on large-scale CCUS-CCS or CC systems, available literature reviews on small-scale CCS for building applications are very limited.

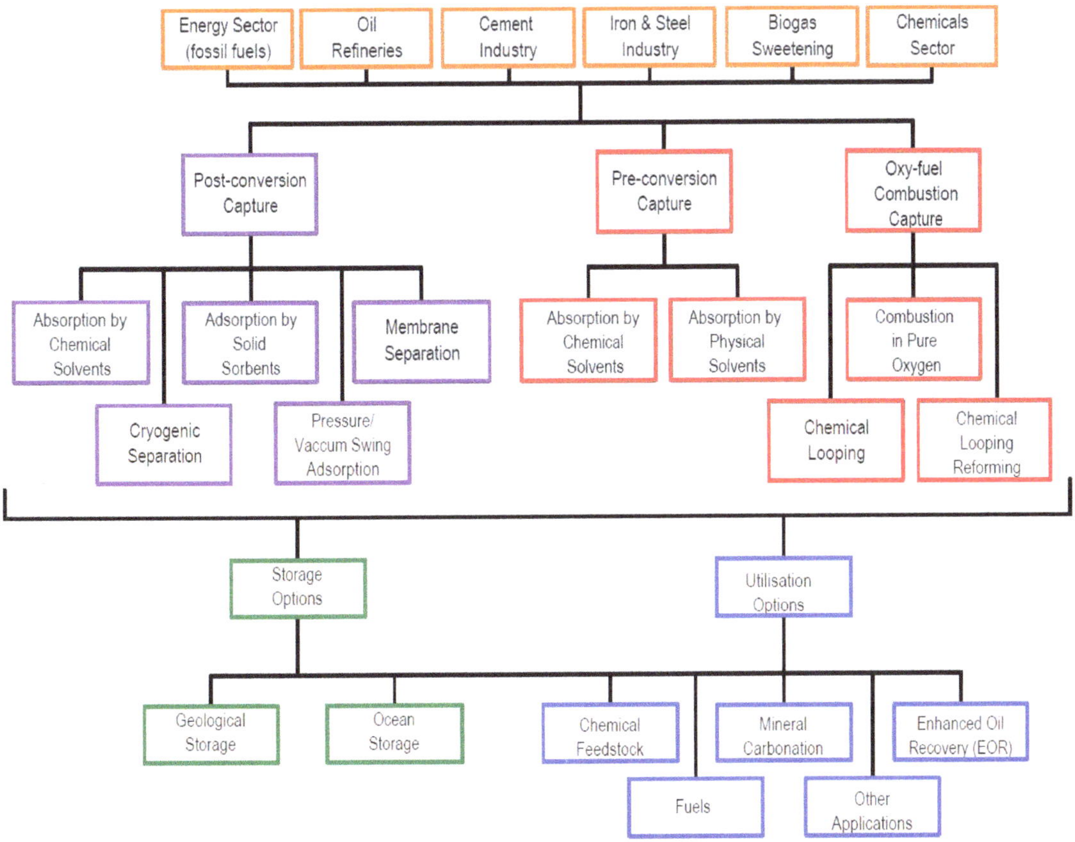

Figure 2. Overview of carbon capture, utilization, and storage options (Adapted from [10]).

Therefore, this review will focus on the current technological developments in carbon capture for decentralized, small-scale CHP generation applications. This is to provide a thorough review of the current technologies in development, and the associated results and challenges, as identified in the literature. For this purpose, the review is structured in the following manner: First the relevant system configurations studied in the literature are presented, and the associated studies introduced (Section 2). Subsequent performance and optimization sections (Sections 3 and 4) then provide detail of the results, challenges, and optimal design parameters from the studies considered. A brief overview for current avenues of storage and utilization is presented, followed by an overview of commercially available small-scale and large-scale technologies (Section 5). Further, general costs of CC systems are briefly reported (Section 6). The next section provides a concise overview of current developments in direct air capture (DAC) technology. Following carbon capture and compression, the next step in the process chain is either storage or utilization (Section 7), while conclusions are drawn in the last section (Section 8).

2. Process and Configurations

This section provides an overview of the micro combined heat and power (micro-CHP) system configurations studied in the literature for CC applications. Each subsection is defined by the prime mover of the energy conversion system. The CC unit must be designed based on exhaust gas CO_2 concentrations and volumetric flow rate and is, therefore, dependent upon the prime mover. The relevant literature associated with each energy system is presented, and the following performance section details results and challenges identified from the same literature.

2.1. Micro-Gas Turbine

The micro gas turbine (mGT) is a suitable technology for small-scale heat and power generation, given the push towards decentralized heat and power production. For decentralized, small-scale generation to be favorable, the solutions must be both carbon free and efficient. Within this framework, mGTs are an attractive solution as they provide high flexibility, as well as a global efficiency of 80% (electrical efficiency of 30% and thermal efficiency of 50%) [59–64]. The range of electrical power output from a typical mGT is between 50–500 kW, which is suitable for small-scale use in multi-family residential, commercial, and institutional applications. The available literature focusses on the Turbec T100 mGT [65], with a nominal power output of 100 kW, as it is well-known and can be considered representative of the current state of the art. The typical CC method uses an absorber-stripper system where the absorbent is a 30 wt% aqueous monoethanolamide (MEA) solution [47,49].

The Turbec T100 is a typical recuperative Brayton cycle mGT [65–67]. Figure 3 shows the schematic of a Turbec T100 mGT with exhaust gas recirculation (EGR) and coupled CC unit. Giorgetti et al. [67] analyzed this system to provide a baseline study on the effects of CC on the mGT cycle. These findings are presented in the performance section of this paper. This models the mGT Turbec T100 at Vrije Universiteit Brussel (VUB) and the Pilot-scale Advanced Capture Technology (PACT) facility at the UK Carbon Capture and Storage Research Centre. Of the literature examined, all studies used this general configuration, although some modified the system to include additional processes and cycles to improve overall performance, as will be discussed.

The general process of the cycle investigated by De Paepe et al. [68] can be described as follows: The air is compressed in a variable speed radial compressor (1) and passed through a recuperator (2) where it is preheated by the exhaust gas arriving from the turbine. The preheated air then enters the combustion chamber, where burning natural gas heats it to a nominal outlet temperature of 950 °C (3). The combustion gas mixture expands across the turbine (4), which delivers the power to drive the compressor and converts the excess power to electricity through a variable speed generator (5). The heat remaining in the gas after the recuperator is recovered through a gas-water heat exchanger (6), which can be used to heat water for combined heat and power systems. The exhaust gases are then split into two streams, one of which is recirculated to the compressor, and the other goes to the CC plant. Exhaust gas recirculation is one of the known emission control technologies for reducing NO_x emissions by recirculating a part of the exhaust gas, while reducing fuel consumption and pumping loss. The EGR ratio is defined by the ratio of intake CO_2 concentration to exhaust CO_2 concentration.

The EGR stream passes through a cooler (7) to maintain high compression efficiency (to be discussed in a later section), and the condensed water is separated (8). The gas is next distributed over a blower (9) to provide a driving pressure increase, followed by a filter (10), which leads to the compressor inlet. A certain ratio of exhaust gases is also directed to the CC plant. The CC plant has two columns, one absorber and one stripper, as shown in Figure 3. A blower (11) provides the required pressure to drive the flue gas in the bottom of the absorber (13). The lean solvent is fed into the top. The interaction of the gas and liquid phases in the absorber drives the CO_2 to the liquid phase, as a result of the concentration gradient at the liquid/gas interface. The rich solvent is then pumped through the heat exchanged (15), where it is heated to a higher temperature by the lean-solvent from the stripper bottom. It then enters the stripper column, where the solvent is regenerated through heat provided by pressurized hot water (16). The vapor at the top of the desorber enters the condenser, in which the water is removed, and nearly pure CO_2 is obtained. The regenerated solvent is pumped back through the rich-lean heat exchanger, and cooled further by an air-cooled plate cooler (18). Wash columns are also mounted at the top of the stripper and absorber columns to eliminate entrained droplets of solvent transported by the flue gas by means of demineralized water. Wash columns have a low energy effect on the CC plant and are neglected in the numerical analysis performed by Giorgetti et al. [66,67].

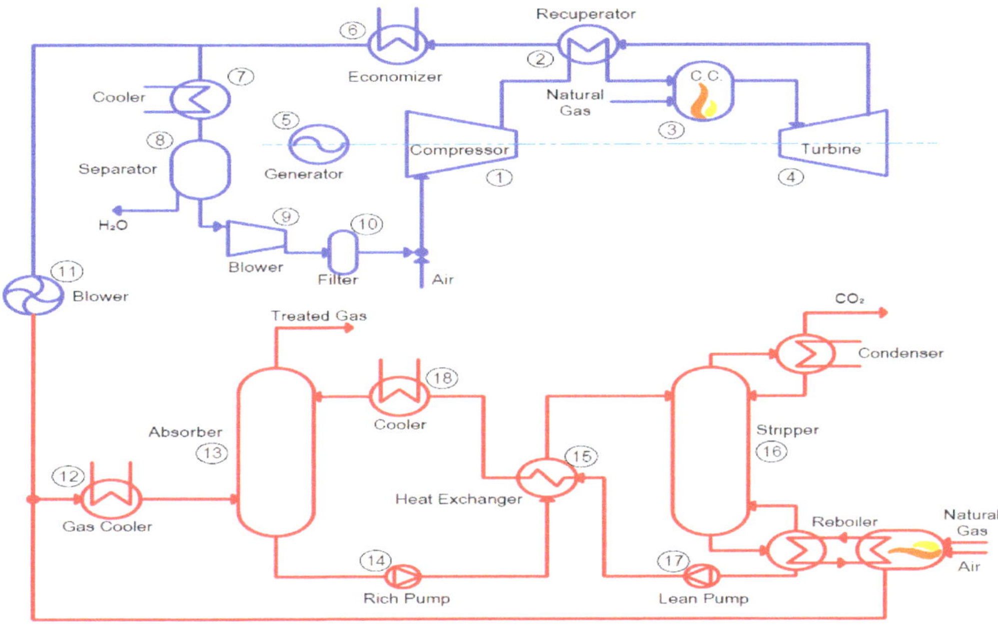

Figure 3. Schematic of the mGT and coupled CC unit representing the mGT Turbec T100 (adapted from [67]).

Table 1 presents the key design conditions of the mGT and CC plant.

Table 1. Key design conditions of the mGT and CC plant (adapted from [67]).

Unit	Parameter	Value
mGT	Electrical power output (kW)	100
	Thermal power output (kW)	153
	Electrical efficiency (%)	30
	Pressure ratio (-)	4.35
	$\%CO_2$ (in the baseline mGT) (%)	1.6
	Turbine outlet temperature (°C)	645
	EGR ratio (-)	0.62
	$\%CO_2$ (in the mGT with EGR) (%)	4.3
Carbon Capture	Absorber dimensions (m × m)	6 × 0.6
	Stripper dimensions (m × m)	6 × 0.45
	Packing Type (-)	IMPT
	Packing Size (mm)	38

This configuration, with the PACT amine capture plant combined with a Turbec T100 mGT is considered in studies conducted by Akram et al. [69], Majoumerd et al. [70] and Ali et al. [71,72]. Their individual findings are detailed in the following performance section of the paper.

Due to the low CO_2 concentration (~1.5 vol%) and high volumetric flowrate of exhaust gases that enter the CC plant, in addition to the substantial residual O_2 amount due to CC plant integration, several cycle modifications have been proposed. One such modification is selective EGR (S-EGR), as opposed to the traditional EGR cycle, as shown in Figure 4. Bellas et al. [73] examined the influences of the S-EGR cycle on the mGT cycle for CC applications. The S-EGR system is similar to the traditional EGR cycle; however, the separated flue gases are passed through a selective membrane system that uses an air sweep stream that blends with the CO_2 passing over the membrane. The CO_2 and air are recirculated to the compressor inlet, while the CO_2 exhausted gases are released to the atmosphere. The parallel and series configurations investigated by Bellas et al. [73] are shown in Figure 4. The parallel configuration was also investigated by Darabkhani et al. [74].

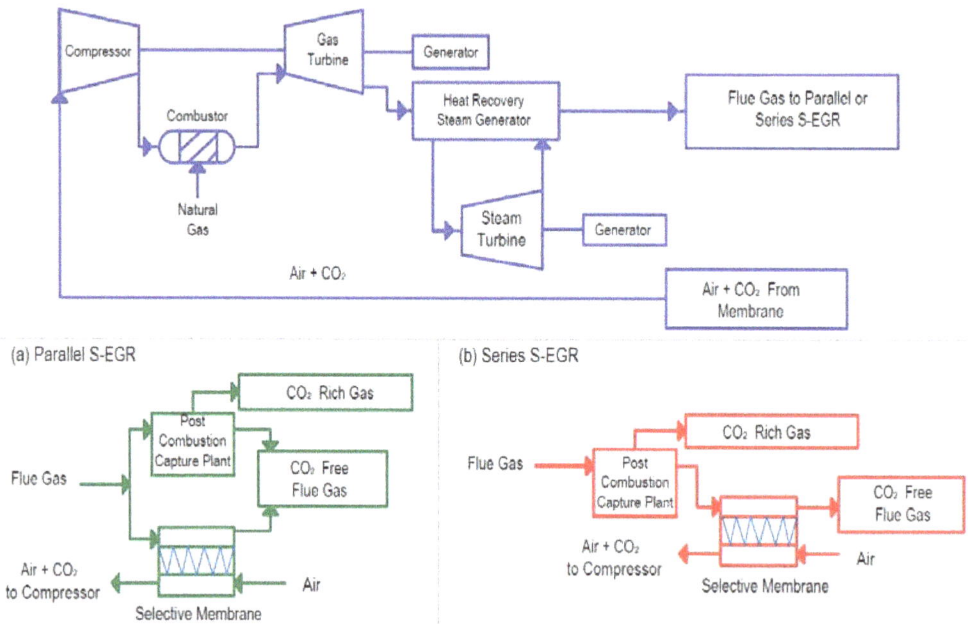

Figure 4. (**a**) Parallel and (**b**) series configurations of the selective EGR cycles (adapted from [73]). (**a**) parallel; (**b**) series.

Several studies are available in the literature that examine the effect of humidification on mGT cycle performance investigated, for instance, by De Paepe et. al. [75,76] and by Best et al. [77]. De Paepe et al. [75] investigated several advanced humidification cycles as applied to mGT. Their results indicated that the REVAP cycle shown in Figure 5 could be considered optimal in terms of waste heat recovery and electrical efficiency. In the REVAP cycle, part of the air/liquid water mixture from injection is employed for compressor aftercooling.

However, the most common humidification cycle studied in the literature is the micro humidified air turbine (mHAT) cycle. Giorgetti et al. [66] assessed the performance of the mHAT cycle through modification of the standard mGT-EGR cycle through the addition of a saturation tower, as illustrated in Figure 6. This cycle can be considered typical throughout the examined literature. The general cycle process is modified as follows: After passing through the compressor the air is humidified at the saturation tower (11). In this configuration, the gas-water heat exchanger (6) is used to heat up the water for the saturation tower, and is no longer, or only partly, used for cogeneration purposes. To balance the water humidification, feedwater enters the circuit (12). A variable pressure pump is also added to drive the circulation of water (13).

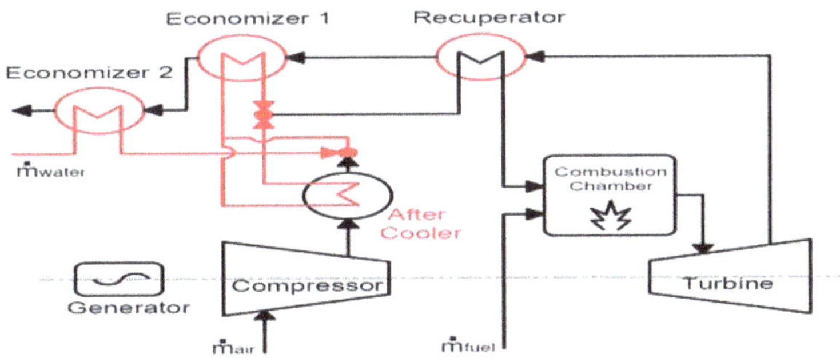

Figure 5. The REVAP cycle, combining liquid water injection with aftercooling and feedwater preheat with economizer 2 (adapted from [75]).

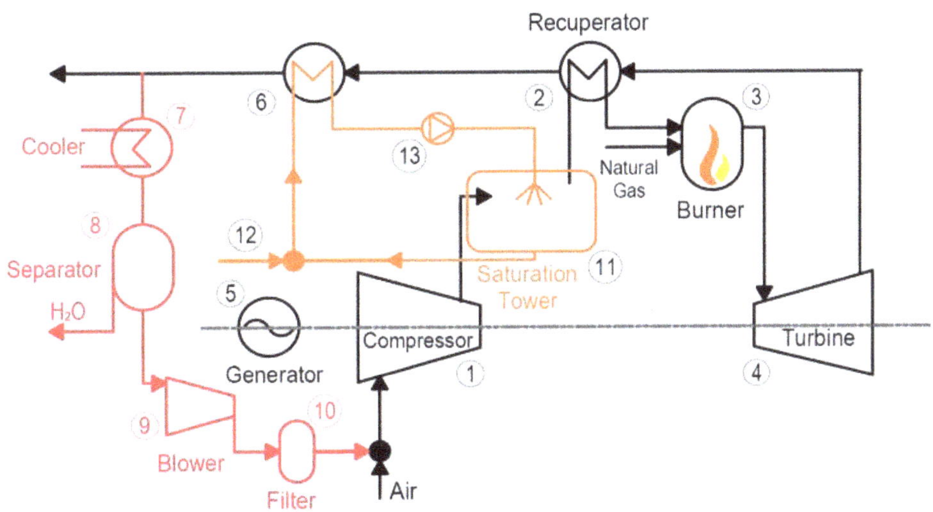

Figure 6. Conventional mGT transformed into a mHAT through the addition of the humidification sub-cycle, depicted in orange (adapted from [66]).

Majoumerd et al. [70] assessed the performance of a typical mGT and mHAT coupled with an advanced post-combustion CO_2 capture unit. The SOA chemical absorption unit also implements monoethanolamine (30 wt% MEA) as a chemical solvent. The capture plant configuration is similar to that studied by Giorgetti et al. [66], but the additional cooler for the lean solvent is removed.

Giorgetti et al. [78,79] investigated the performance of an mHAT system coupled with an amine-based absorption plant with energy integration between the two plants, which is not considered in other studies available in the literature. The energy integration between the CC plant and mGT system is presented in Figure 7. Waste heat from the exhaust gases is utilized based on the heat demand, which is represented by activation of blocks A, B, C, or D, as discussed in detail in Section 4.1.4.

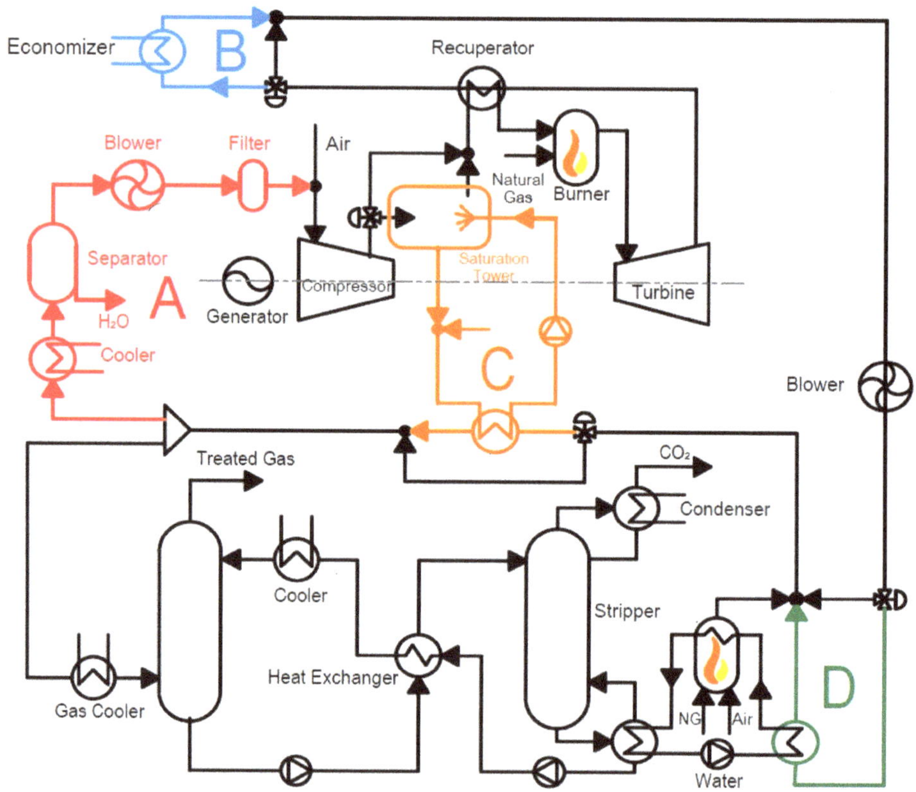

Figure 7. Plant layout displaying the standard EGR in block A, the CHP application in block B, and, in the case of no heat demand where block B is bypassed, block C for humidification and block D for reboiler duty are activated (adapted from [78]).

2.2. Hybrid Fuel Cell Systems

Fuel cells such as those in small-scale applications [80,81] or in micro-scale applications [82,83], have high efficiency, as electricity is generated through an electrochemical reaction as opposed to a series of energy conversions. In the literature, hybrid fuel cell systems on a small-scale are scarce, while hybrid systems with large-scale gas turbines are more readily available, such as hybrid solid oxide fuel cells directly coupled to a gas turbine [84,85], or indirectly coupled to a gas turbine [86] in power plants, and hybrid solid oxide fuel cell-gas turbine cycles using alternative fuels [87].

Roohani Isfahani and Sedaghat [88] developed a novel system with a unique combination of a solid-state fuel cell, micro gas turbine, and CC unit to use natural gas energy in a more effective approach. The fuel cell power output considered was between 950 and 1360 kW, making the system applicable to decentralized energy generation with application to multi-use commercial buildings, as opposed to single-family dwellings and small businesses. The system comprises three reactors for splitting hydrogen and carbon dioxide from natural gas through the three-reactor chemical looping hydrogen generation (TRCL). Several other studies are available that focus on hybrid power plants of solid oxide fuel cell (SOFC), and micro-GTs operating at baseload [89,90] or part-load [91], applying various operating strategies [92,93] or using alternative fuels [94] for micro-CHP applications. However, Roohani Isfahani and Sedaghat [88] extended the above systems to integrate with a CC plant and TRCL system, which is pertinent to the topic of this review. The hybrid power plant is made up of a reformer, three-reactor chemical looping for hydrogen production, a fuel cell, micro-GT, an internal heat exchanger, and the CO_2 capture loop. Natural gas is broken down into carbon dioxide and hydrogen constituents in the reformer, which is then passed into the TRCL reactors. The TRCL comprises of three reactors: fuel, steam and air reactors. In the first stage, fuel is injected into the fuel reactor (FR). In this reactor, hematite (Fe_2O_3), which contains a significant amount of elemental oxygen, is mainly reduced to FeO (Wüstite, a mineral form of iron (II) oxide) at 950 °C at the upper riser, and 890 °C at the bubbling bed. The fuel is converted into CO_2 and H_2O. In this stage, the CO_2 is ready to be absorbed as a product. In the second reactor (steam reactor or SR) FeO (Wüstite) reacts exothermically with steam and creates magnetite (Fe_3O_4) and hydrogen at 950 °C. In the air reactor (AR), entered magnetite (Fe3O4) reacts exothermically with pure air, to produce hematite (Fe2O3) and oxygen-depleted air as products. The overall reaction changes methane (CH_4) into hydrogen and CO_2. CH_4 is also converted to H_2 and CO_2. Therefore, CO_2 is integrally separated from fuel. The TRCL reactors are operated at a pressure of 3 bar.

Steam from the heat recovery steam generator reacts with fuel in the reformer. The H_2 from the steam reactor is then fed into the anode of the fuel cell, and O_2 depleted air from the air reactor goes to the cathode to start electricity production. The fuel reactor exhaust CO_2 is fed to the CO_2 capture loop. Unreacted hydrogen from the SOFC is fed into the combustor and burned to produce sufficient hot gas to run the micro gas turbine. This latter produces energy that runs the adjacent compressor to compress the inlet air to the air reactor. The turbine exhaust gases are also used to preheat the incoming natural gas. In the CO_2 capture loop, the incoming CO_2 from the fuel reactor of the TRCL is cooled in the CO_2 heat recovery steam generator (CO_2 HRSG). The CO_2 undergoes several compression and cooling processes to reach the pressure required in the pipeline. Figure 8 presents the overall system schematic, and Figure 9 depicts a detailed schematic of the TRCL.

2.3. Biomass-Fired Organic Rankine Cycle

The organic Rankine cycle (ORC) system is an attractive technology for cogeneration applications in the 200–1500 kW range, mainly used in waste heat recovery [95–97], at small-scale for industrial or commercial buildings [98,99], in small-scale and micro-scale biomass fueled CHP systems [100], or at micro-scale for residential applications [101]. However, few applications with small-scale ORCs with integrated CCS are commercially available. Zhu et al. [102] designed a biomass-fired ORC for small-scale CHP systems, coupled with an MEA-based CC unit, for which they performed an extensive thermo-economic simulation study. The system implements biomass combustion as the primary energy source and the ORC as secondary. The electric power output for the considered system ranges from 100–500 kW and is therefore applicable for small-scale applications, such as residential and commercial buildings, office building blocks, etc. The system schematic is provided in Figure 10. The process is defined in the following manner: The biomass fuel is combusted in the biomass boiler and heat is transferred to the pressurized hot water during process 9–8. The heat is then absorbed in the evaporator (4–1) by the organic working fluid, which then passes over the expander to produce power (1–2). The working fluid then enters the condenser where it is cooled by the cooling water (2–3). The cooling water absorbs the waste heat (10–11) to generate domestic hot water or discharges the heat in the cooling towers (17–18). From the biomass boiler, the flue gases preheat the combustion air entering the boiler (7–16), and then enter the MEA chemical absorption unit. Section 3.3 details the results and challenges encountered by Zhu et al. [102] as they relate to the system performance.

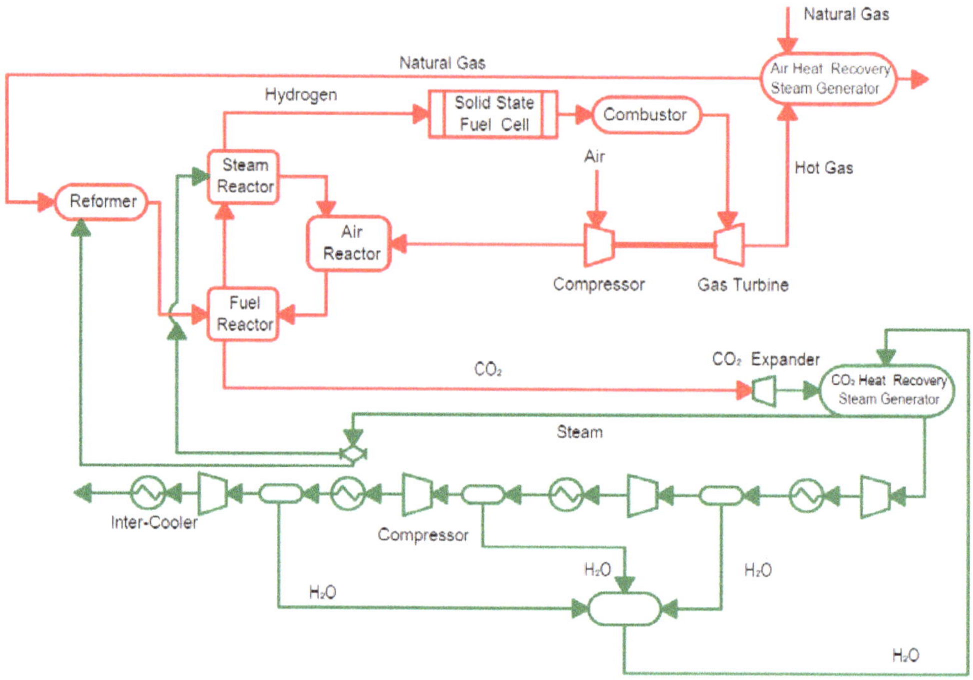

Figure 8. Hybrid power plant schematic (adapted from [88]).

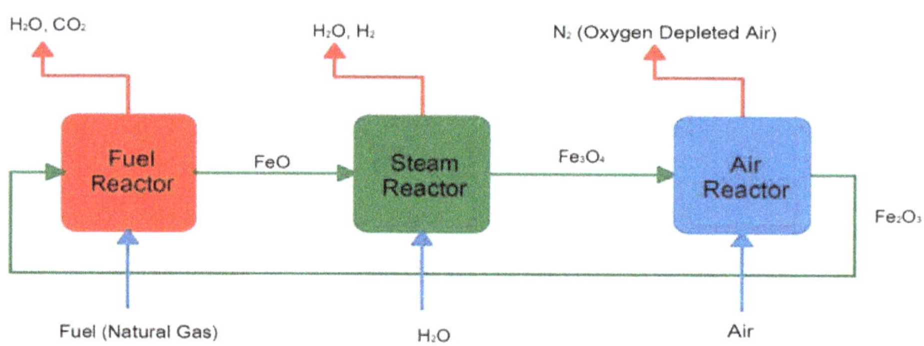

Figure 9. Conceptual design of three-reactors chemical looping for H_2 generation (adapted from [88]).

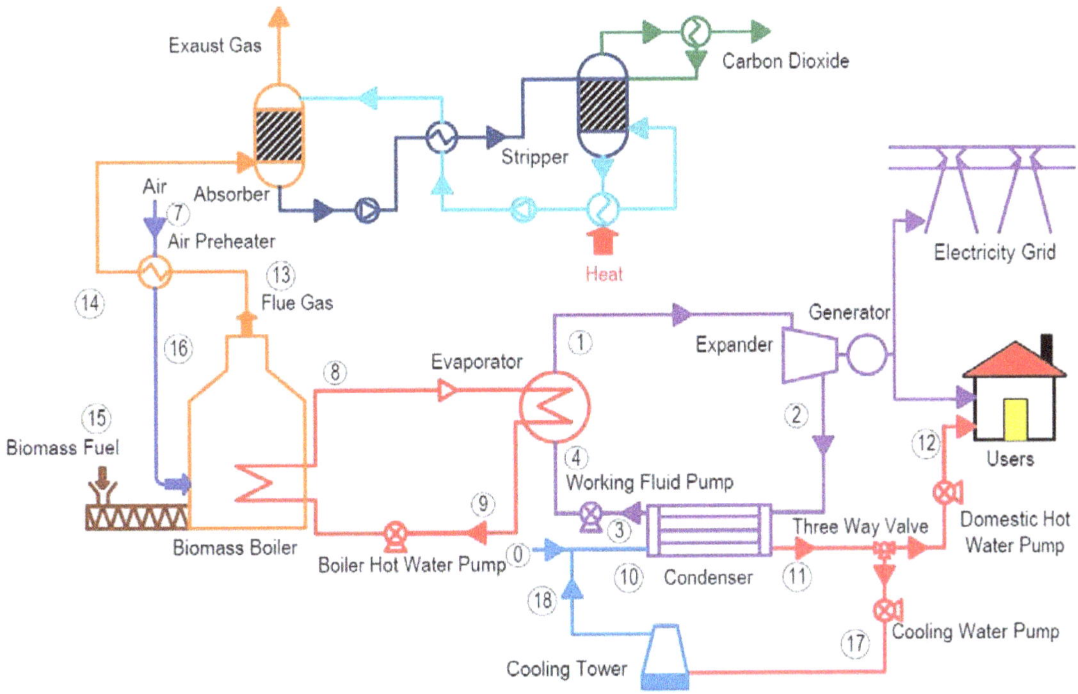

Figure 10. Schematic process flow diagram of biomass-fired ORC-CHP system with MEA-based CC (adapted from [102]).

3. Performance

This section details the specific results and challenges pertaining to the performance of each configuration from the selected studies introduced in Section 2. It is divided into subsections based on the prime mover in the energy conversion system.

3.1. Micro Gas Turbine

Micro gas turbines offer the lowest CO_2 emissions per produced kW, making them an attractive option for small-scale CHP energy generation systems. However, the carbon dioxide emissions must be captured in order to achieve carbon neutral energy production. Challenges encountered when integrating carbon capture technology into mGT energy conversion cycles result most notably from the low CO_2 concentration and high volumetric flow rate of exhaust gases (~1.5 vol%), plus the significant remaining O_2 concentration that leads to solvent degradation. These factors have a negative impact on the size, energy use and economic performance of the downstream CC plant, while also resulting in solvent degeneration if an amine-based CC plant is utilized. In order to minimize these effects, namely the cost of the energy penalty, technologies such as auxiliary firing, exhaust gas recirculation (EGR), selective exhaust gas recirculation (S-EGR), humidified cycles, and oxy-fired gas turbines cycles have been proposed. Supplementary firing and oxy-fired GT cycles have not been explored on the small-scale in the literature; therefore, the following subsections detail the overall impact of the CC plant, followed by the impact of EGR, S-EGR, and humidification on cycle performance.

3.1.1. Impact of Carbon Capture

As previously described, integration of a CC plant with an mGT results in a high energy penalty. Giorgetti et al. [67] assessed the effect of a CC plant on the global performance of the mGT through numerical simulations in Aspen Plus. Their results indicated that the cycle performance was greatly impacted by the thermal energy demand for the stripping process (reboiler duty), reducing the total electric efficiency by about 6.2 absolute percentage points. Table 2 presents the electrical efficiency of the entire plant compared to that of the traditional mGT for various part loads up to nominal load (100 kW). Further, the impact of varying the electric power output of the mGT is examined in terms of reboiler duty and global efficiency. Table 3 shows that the reboiler duty had a positive near-linear function with the power yield. While the results showed an increase in reboiler duty with increasing load output, the specific reboiler duty remained relatively constant, with a lower range of 4.25 MJ/kgCO_2 at 75 kW and an upper range of 4.38 MJ/kgCO_2 at nominal load. It is important to mention that the simulations implemented an optimized CC plant, with a liquid/gas (L/G) ratio of 1.2, as well as an EGR ratio of 0.61 and 90% CO_2 removal rate.

Table 2. Comparison of the global plant efficiency with EGR and CC and traditional mGT efficiency (adapted from [67]).

Parameter	Values			
Electric power output (kW)	75	85	95	100
mGT global plant efficiency (%)	28.0	28.3	28.5	29.0
mGT + EGR + CC global plant efficiency (%)	22.0	22.2	22.2	22.2

Table 3. Effect of electric power output on reboiler duty (adapted from [67]).

Parameter	Values					
Electric power output (kW)	75	80	85	90	95	100
Reboiler duty (kW)	58	61	65	68	72	75

It is also important to point out the influence of ambient air temperature on electrical efficiency. As ambient air temperature rises, the air density reduces, causing a lower mass flow of air within the engine and reduced power production. Consequently, a higher heat input is required to increase the air and fuel mass flow to generate the nominal power output, resulting in efficiency decrease. The ambient air temperature also affects the oxygen concentration at the combustor inlet; however, these effects are marginal and are not of concern [70].

3.1.2. Exhaust Gas Recirculation

Exhaust gas recirculation (EGR) is a proposed technology to decrease the cost of the energy penalty from CC resulting from the low CO_2 content of exhaust gases. It is worth noting that post-combustion CC is more efficient for large-scale GTs where CO_2 concentrations are higher (~3.8 to 4.4 mol%) than for mGTs (~1.6 to 1.8 mol%) [71] Exhaust gas recirculation supplies three main benefits—it increases the exhaust gas CO_2 concentration for a reduced carbon capture energy penalty, it reduces harmful NO_x emissions, and it decreases the volumetric flow rate to the CC plant by recirculating a fraction of the exhaust gases back to the compressor inlet [68]. Akram et al. [69] experimentally determined that, per unit percentage rise in CO_2 concentration, the specific reboiler duty decreased by around 7.1%, and numerically predicted a 6.6% reduction.

However, EGR also introduces some challenges. The combustion stability diminishes and unburned emissions increase as the O_2 concentration at the combustor inlet is decreased; electrical efficiency is decreased through auxiliary energy losses to the fan that drives the recirculated gas, and the compressor inlet temperature is increased, resulting in a slight decrease in thermodynamic performance [103]. Giorgetti et al. [66,103] found that the EGR blower consumed 4.5 kW of power at nominal operation conditions accounting for 4.5% of the electrical power output. Removing this from the efficiency calculation resulted in nearly identical efficiency relative to the traditional mGT cycle. The remaining difference was small due to the change in the inlet mixture temperature with EGR.

Majoumerd et al. [76] found that with 40% EGR the CO_2 content of the exhaust gas could be increased from 2 mol% to 3.4 mol%, through simulations with a validated thermodynamic model. This represents a 67% increase relative to the reference mGT cycle. Similarly, through simulations using Aspen Hysys and IPSEpro, Ali et al. [71,72] established that the CO_2 concentration in the exhaust gas of the mGT with 55% EGR was 2.2 times greater than the traditional mGT cycle, where baseline CO_2 content in the exhaust was 1.46 mol%. This increase resulted in a 40% reduction in specific reboiler duty, demonstrating the advantages of EGR for CC applications, owing to the reduction in cost from smaller absorption/stripping columns and the reboiler.

Majoumerd et al. [70] determined that while the CO_2 concentration in flue gases was increased with EGR, the global efficiency was decreased from 23.0% without EGR to 22.5% with EGR. Ali et al. [71,72] came to the same conclusion, that electrical efficiency of the mGT cycle is decreased by EGR. Their results showed a decrease in mGT electrical efficiency from 32.1% to 29% at a 55% EGR ratio, when the effects of the CC plant were not considered. As discussed, this can be attributed to the blower power required for recirculation and the changes in the fluid thermodynamic properties that effect compressor and turbine operation. However, the global efficiency reduction due to EGR is small, and the literature appears unanimous in the conclusion that EGR is desirable due to the cost decrease for CO_2 capture.

Best et al. [104] experimentally assessed the effects of EGR on mGTs through CO_2 injection in a Turbec T100. Their results showed that, at low power outputs (50 kW) with 125 kg/s CO_2 augmentation, CO emissions increased by 109% and unburnt CH_4 emissions by 338%. However, they concluded that emissions were not significantly impacted at higher load factors. Further, due to the lower combustion temperatures, NO_x emissions showed a decreasing trend with CO_2 enhancement.

3.1.3. Selective Exhaust Gas Recirculation

As discussed, EGR increases the CO_2 concentrations of exhaust gases while decreasing the volumetric flow rate to the CC plant by recirculating a portion of the exhaust back to the compressor inlet. This has been demonstrated to reduce the energy penalties of carbon capture, as well as to decrease the capital cost of the system as a result of the reduced system size. However, increasing the fraction of flue gas to be recirculated decreases the oxygen concentration at the combustor inlet, resulting in flame instabilities and decreased combustion efficiency, and high CO and unburned hydrocarbon (UHC) emissions. Therefore, the EGR ratio is limited by an optimal O_2 concentration of 16 vol% at the combustor inlet. As presented in Section 2.1 of this review, S-EGR is a proposed cycle modification to increase CO_2 content of exhaust gases to a higher degree than achievable with EGR, without compromising O_2 concentrations at the combustor inlet required for combustion stability.

Darabkhani et al. [74] studied the performance of a parallel S-EGR configuration through both simulation and experimental testing. The focus of the study was on the performance of a commercially available, polydimethylsiloxane organic polymer membrane (purchased from PermSelect Ltd., Ann Arbor, MI, USA), as investigated both experimentally and through simulation. Through process simulations, it was found that CO_2 concentrations could be achieved of up to 14.9% with 60% EGR, with a 90% CO_2 removal rate from the membranes.

Challenges encountered with S-EGR are flame instability and combustion efficiency leading to increased CO and UHC emissions due to the higher CO_2 content achieved in the flue gases. Bellas et al. [73] performed experimental studies on the Turbec T100 mGT with the goal of investigating the effects of CO_2 enrichment on the performance of an mGT. To emulate the effects of S-EGR, CO_2 was injected into the compressor inlet. Injection rates of 0 to 300 kg/h CO_2 (1.7 to 8.4 vol% at 100 kW and 1.4 to 10.1 vol% at 60 kW) were considered, and the effects on gas turbine performance were assessed. This represents a nearly six-fold increase in CO_2 concentration, which is typical of S-EGR. It was found that high levels of CO_2 injection modified the specific heat capacity and density of the oxidizer, decreasing the engine speed and system temperatures. The CO and UHC emissions increased greatly at part loads, whereas at nominal load they experienced little change with increased injection rates. This is an effect of incomplete combustion caused by poor fuel and air mixing, inadequate flame stability, and lesser combustion temperatures. At the highest injection rates (300 kg/h of CO_2), the NO_x emissions were lower than at the baseline (no injection). This is a result of lower combustion temperatures at increased CO_2 content.

The influence of different S-EGR injection rates on flue gas emissions across the 60–100 kW operating envelope are shown in Table 4.

Table 4. Effect of CO_2 injection rates for the given operating envelope (60–100 kW) on CO and UHC emission concentrations (adapted from [73]).

Parameter			**Values**						
CO_2 Injection Rate (kg/h)			**0**	**50**	**100**	**150**	**200**	**250**	**300**
CO concentration (ppm dry)	Operating power (kW)	60	100	140	180	220	300	420	510
		70	20	20	50	80	100	220	305
		80	0	0	0	0	30	70	100
		90	0	0	0	0	20	50	50
		100	0	0	0	0	0	10	20
UHC ($CH_4 + C_2H_6$) (ppm dry)		60	8	10	20	30	40	78	105
		70	0	0	0	5	10	35	50
		80	0	0	0	0	0	5	10
		90	0	0	0	0	0	5	5
		100	0	0	0	0	0	0	0

3.1.4. Humidification

As discussed in Section 3.1.2, EGR enhances the efficiency of the CC plant, but decreases the electrical efficiency of the mGT cycle resulting in a marginal decrease in global efficiency. Humidification is a proposed method for improving mGT cycle efficiency. In general, the overall economic performance of the mGT powered mCHP system can be improved through improving the electric efficiency of the mGT at the time of low heat demand.

Studies are available that assess the impact of humidification on the global efficiency of the mGT and coupled CC plant, as well as studies that examine the impact on individual mGT component performance. Both aspects are reviewed in the following discussion.

Giorgetti et al. [66] found that humidification of the traditional mGT cycle can completely offset the efficiency losses introduced by the EGR energy penalty. Table 5 provides the simulation results from the standard mGT with EGR and the mHAT with EGR, both integrated with an amine plant and operating at 100 kW with 16% O_2 in the combustor inlet. Of note is that the EGR ratio is maximized based on the limit introduced by the minimum 16 vol% oxygen content at the combustor inlet, which is discussed further in Section 4.1 of this review. The findings show that water injection decreases the oxygen concentration of the mixture. Hence, the maximum EGR ratio to maintain 16 vol% oxygen at the combustor inlet is lower for mHAT than the traditional mGT cycle, resulting in a larger diameter absorber column and increased cost. The authors also found that humidification significantly improved global cycle efficiency, increasing it from 21.3% with the traditional mGT to 23.9% with mHAT [66].

Table 5. Comparison of mHAT and mGT at 100 kW electrical load, 16% O_2 in the combustor inlet, and constant turbine outlet temperature (adapted from [66]).

Parameter	mGT	mHAT
EGR_{ratio} (-)	0.61	0.46
Mass flow rate of the exhaust gas (kg/s)	0.295	0.360
CO_2 concentration in the exhaust gas (%)	4.3	3.0
Absolute electrical efficiency decrease (%)	−1.3	−0.9

Similar results were presented by Majoumerd et al. [70] through simulations using a validated thermodynamic model. The results showed significantly increased performance for the mHAT cycle compared to the traditional mGT, with 25.8% cycle efficiency compared to 23.0% and 22.5% efficiency for the baseline mGT and mGT-EGR cycles, respectively. Of note is that the cycle efficiency includes the effects of the coupled CC plant.

Carerro et al. [105] experimentally studied the effects of water injection with a saturation tower on the mGT cycle. They found that, overall, the electrical efficiency of the humidified cycle increased up to 4.2 absolute percentage points, similar to that found numerically.

Therefore, it can be concluded that global cycle efficiency is improved through humidification, and entirely compensates for energy losses from EGR. It is important to mention that this result is independent of energy integration between the CC unit and mGT/mHAT, in which waste heat is leveraged to decrease the thermal energy requirement of the stripper. This is discussed further in the optimization section of this review, in which Giorgetti et al. [66] assess a mGT/mHAT coupled with a CC unit with energy integration between the systems.

3.2. Hybrid Fuel Cell Systems

Roohani Isfahani and Sedaghat [88] evaluated the performance of the hybrid system of a solid-state fuel cell and micro gas turbine with a coupled carbon capture plant, as introduced in Section 2.2. A major benefit of this system is the 100% capture ratio achieved by compressing CO_2 to the liquid state, compared to the 90% CO_2 capture common in the micro gas turbine power generation systems. The influence of plant pressure and SOFC temperature on the system effectiveness and performance has been investigated in detail. The system efficiency can be maximized by increasing the SOFC temperature and operating pressure, although the effect of working temperature clearly has less influence than that of the pressure. It was found that at an SOFC temperature of 1000 °C, an SOFC pressure of 17.5 bar, and a fuel utilization factor of 0.8, the global efficiency reached 48.3% natural gas LHV (lower heating value). It is also important to note that SOFC operating temperatures may not exceed 1000 °C, and that while increasing pressure increases efficiency, it simultaneously increases capital cost, which must be taken into consideration. Moreover, the effect of the utilization factor and plant pressure on power plant efficiency and fuel cell power output has been assessed. The utilization factor may be expressed as the ratio of consumed fuel in fuel cell stacks and the total quantity of fuel introduced to any type of fuel cell and has a typical range of 0.75–0.9. If it is increased, the power output of the fuel cell is increased while the generated power of the gas turbine is decreased. However, given that fuel cells are more efficient than gas turbines, more fuel is utilized in a more effective manner. Therefore, increasing the utilization factor increases the overall efficiency. Further, the maximum net efficiency was found to be 51.4% LHV at an exhaust pressure of 2.5 bar, temperature of 1000 °C, plant pressure of 22.5 bar, fuel utilization factor of 0.9, and CO_2 expanded exhaust pressure of 2.8 bar.

3.3. Organic Rankine Cycle

The biomass-fired organic Rankine cycle is a suitable technology for distributed CHP. Although the ORC is an established technology for cogeneration across the range of 200–1500 kW, few are commercially available on the small- and micro-scale, where the electrical output ranges from 100 to 500 kW. Zhu et al. [102] investigated the ORC-based biomass fueled micro-CHP system with integrated MEA-based CC to assess the thermodynamic and economic performance of eleven working fluids. From the perspective of the feasibility for distributed energy generation below 500 kW, the net power outputs ranged from 175.49 kW for isobutene to 413.82 kW for cyclopentane. Of all the working fluids considered, it was found that cyclopentane had the best thermodynamic performance with a power efficiency of 13.70% and exergy efficiency of 16.21%. This was followed by R141b, R113, R123 and pentane. It is important to note that the reported values in the thermodynamic analysis did not include CO_2 capture. However, from the economic assessment, HFE7000 had the largest net present value (NPV) of 2052.42 × 10^4 USD and the highest profit ratio of investment (PRI) of 5.45, followed by R1233zd-I, isobutane, isopentane and R113. The results of the economic analysis performed by the authors [102] is detailed in Section 6 of this review. This takes into account the thermo-economic aspect of the system with and without CC and draws associated conclusions. In general, the authors deduced that the biomass-fired ORC system for micro-CHP with optimized parameters is economically significant and sustainable, and allows for negative CO_2 emissions when integrated with CC.

4. Design Optimization

This section will focus on design optimization of the configurations presented in Section 2 and detailed in Section 3. A myriad of studies are available pertaining to the micro gas turbine in the context of small-scale CC, as well as the optimization of various system parameters and design considerations for optimal system performance. However, studies on the hybrid solid state fuel cell and ORC cycles with CC, in a small-scale context, are limited and therefore not included in this section of the review.

4.1. Micro Gas Turbine

Numerous studies are available that assess mGT cycle modifications for design optimization, such as EGR, S-EGR, and humidification, and their impact on individual mGT components in the context of integration with a CC plant. The following section focuses on studies that seek to optimize both the coupled CC plant and the modifications to the mGT cycle for CC applications. In the final subsection, a new plant layout for improved energy integration will be investigated to maximize global efficiency.

4.1.1. Exhaust Gas Recirculation

As mentioned previously, the EGR ratio is limited by the oxygen concentration at the combustor inlet. Some researchers have experimentally studied the effects of the oxygen content of the oxidizer on flame stability. Their results showed that UHC and CO emissions showed large increases below 16 vol%, and flame stability was compromised below 14 vol%; there was stable combustion with 16–18 vol% oxygen in the oxidizer. Therefore, optimization of the EGR ratio must be performed with consideration of the oxygen content at the combustor inlet.

Majoumerd et al. [70] utilized an experimentally validated thermodynamic model to assess the effect of EGR on flue gas component concentrations. They found that at 40% EGR, the concentration of oxygen in the oxidizer decreased to 18.7 mol%.

Ali et al. [71] performed numerical simulations to maximize the EGR ratio; the results are presented in Table 6. They determined that the EGR ratio must be kept equal to or below 0.55 to maintain 16 vol% O_2 concentration at the combustor inlet. This follows from CO_2 enrichment of 1.6 mol% to 3.7 mol% in the flue gases.

Table 6. Influence of EGR ratio on O_2 molar fraction at the combustor inlet and outlet, and on the CO_2 and O_2 molar fraction in flue gas (adapted from [71]).

Parameter	Value				
EGR (%)	0	0.2	0.4	0.6	0.8
O_2 at combustor inlet (%)	21.0	20.1	19.0	16.0	9.0
O_2 at combustor outlet (%)	17.1	16.0	15.0	12.0	7.0
O_2 in flue gas (%)	17.8	16.2	15.1	12.9	5.1
CO_2 in flue gas (%)	1.9	2.75	3.0	4.0	8.0

4.1.2. Humidification

De Paepe et al. [75] assessed the performance of several advanced humidification cycles through numerical simulations to determine the impact on turbine electrical efficiency, emissions, flue gas mixture concentrations, and overall thermodynamic performance. The following advanced humidification cycles were assessed: compressor inlet air cooling, water atomizing inlet air cooling (WAC), regenerative evaporation (REVAP), steam injected gas turbine (STIG) cycle, mHAT, and mHAT+. Based on the assessment, it was determined that the REVAP cycle was the most efficient solution following from its ability to recover additional waste heat from the evaporation heat from the condensing water in the stack. Applying the REVAP cycle to the traditional dry mGT cycle resulted in an efficiency of 37.1% at nominal load, an increase of 4.3 percentage points. Although the REVAP cycle was optimal in terms of efficiency, the cycle layout is far more complicated compared to the standard direct preheated water injection, and the efficiency augmentation is quite insufficient. It is of note that the efficiency of the standard mHAT cycle at nominal loading is 36.1%.

4.1.3. Carbon Capture Plant

As discussed in Section 2.1, a standard aqueous solution of monoethanolamine (with 30 wt% MEA) is the chemical solvent used in every study considered. The regeneration temperature would be set at 120 degrees to avoid thermal degradation of the MEA solvent and corrosive effects [70].

Agbonghae et al. [106] performed simulations using a validated model of the Turbec T100 mGT integrated with an amine-based (MEA) chemical absorption plant in Aspen Plus. The optimum liquid/gas (L/G) ratios and the lean CO_2 loading for 90% capture is reported in Table 7 for flue gases of 3 to 8 mol% CO_2 content. Optimum L/G ratios result in relatively similar specific reboiler duty values, as seen in Table 7. However, the variations in the L/G ratio effects the optimal height for the absorber and stripper columns and must be considered for capital costs and operating costs.

Table 7. Optimal liquid/gas ratio, lean loading, and specific reboiler duty for flue gas with CO_2 compositions from 3 to 8 mol% (adapted from [106]).

Parameter	Value					
Flue gas CO_2 concentration	3	4	5	6	7	8
30 wt% MEA solution (wt%)						
L/G (-)	0.907	1.253	1.496	1.984	2.220	2.454
Lean CO_2 loading (-)	0.247	0.255	0.249	0.265	0.259	0.254
Specific reboiller duty (MJ/kg CO_2)	4.399	4.390	4.374	4.441	4.431	4.437
35 wt% MEA solution (wt%)						
L/G (-)	0.806	1.003	1.247	1.488	1.973	2.454
Lean CO_2 loading (-)	0.260	0.250	0.250	0.250	0.273	0.285
Specific reboiller duty (MJ/kg CO_2)	4.236	4.261	4.266	4.282	4.245	4.298
40 wt% MEA solution (wt%)						
L/G (-)	0.806	1.003	1.247	1.488	1.973	2.454
Lean CO_2 loading (-)	0.287	0.278	0.278	0.278	0.298	0.309
Specific reboiller duty (MJ/kg CO_2)	4.114	4.112	4.115	4.125	4.138	4.200

4.1.4. Energy Integration

When operating in full CHP mode, such that the waste heat is used for water heating purposes, mGT efficiency is around 80% [61–63,68]. However, when the thermal demand is zero and the waste heat is not used for external heating, mGT efficiency is decreased to purely electrical efficiency. Therefore, the economic feasibility of the mGT integrated with carbon capture is severely diminished. The aforementioned studies all address diverse characteristics of the EGR cycle and humidification of mGTs, with a focus on individual component performance, without consideration of the most effective use of waste heat. To fill this gap, Giorgetti et al. [78,79] investigated the energy integration and strategy optimization of an mGT/mHAT in combined heat and power mode, when coupled with a CC unit. This addressed cycle optimization for waste heat recovery based on varying thermal demands. The system schematic is shown in Section 2.1, Figure 7.

When operating in full CHP mode, the Turbec T100 has an electrical efficiency of 29.4% at 100 kW, and a thermal efficiency of 49.5%, providing 166 kW of thermal power. This results in a total efficiency of 79%. The total efficiency is slightly decreased, by 1.3 absolute percentage points, when EGR is applied. When the capture plant is added, the global efficiency is reduced to 59.2%. The results are summarized in Table 8.

Table 8. Comparison of the electrical efficiencies for various plant layouts functioning in CHP mode (adapted from [78]).

Parameter	mGT	mGT + EGR	mGT + EGR + CC
Blocks	A	A + B	A + B
η_{el} (%)	29.4	28.1	21.4
η_{th} (%)	49.5	49.5	37.8
η_{tot} (%)	78.9	77.6	59.2
$\Delta\eta_{el}$ (compared to mGT)	-	−1.3	−8.0

When the thermal demand is zero, the overall efficiency is decreased to the electric efficiency. The waste heat can be used either for mGT humidification to increase electrical efficiency or can be employed to reduce the thermal demand of the stripper reboiler to reduce the energy penalty. Several plant layouts based on Figure 7 in Section 2.1 are examined, and the electrical efficiency noted. These are summarized in Table 9. The solution with the lowest energy penalty (2.6% absolute percentage points compared to the reference case) is the mHAT with EGR, heat recovery, and CC unit. This achieves an electrical efficiency of 26.8%.

Table 9. Comparison of the electrical efficiencies for different plant layouts where there is no heat demand (adapted from [78]).

Plant Layouts	Blocks	η_{el} (%)	$\Delta\eta_{el}$ (Compared to mGT) (%)
mGT	-	29.4	-
mGT + EGR + CC	A	21.4	−8.0
mHAT	C	32.4	+3%
mHAT + EGR	A + C	31.5	+2.1
mHAT + EGR + CC	A + C	24.1	−5.3%
mGT + EGR + CC + Heat Recovery	A + D	25.9	−3.5%
mHAT + EGR + CC + Heat Recovery	A + C + D	26.8	−2.6%

Moreover, Giorgetti et al. [79] extended the previous study to include part thermal loading. As determined in the study investigated in [78], dry operation of the turbine where the entire thermal output (165 kW) is used for external heating purposes provides the highest overall efficiency, identified in this study as 59.6% (case (a) in Table 10). When the thermal demand decreases from full CHP mode, a threshold exists where the reboiler duty for the stripping process is completely satisfied by waste heat from the mGT cycle (case (b) in Table 10). After CC heat recovery, the exhaust gas has remaining thermal power of 90 kW for cogeneration purposes, resulting in a total system efficiency of 54.7%. In the third case investigated, the mHAT is applied and waste heat is used partially for humidification purposes, and part for CC energy recovery (case (c) in Table 10). However, the external boiler is still required to satisfy the thermal demands of the stripping process. The authors concluded that the overall electrical efficiency when using the energy integration methods was higher for dry operation than humidified operation, due to energy degradation of the flue gases. This result is different than that found in previous studies performing component-wise optimization. Table 10 compares the efficiencies of different plant layouts.

Table 10. Efficiencies of plant layouts (adapted from [79]).

Parameter	mGT Case (a)	mGT Case (b)	mHAT Case (c)
Electrical power production (kW)	96	96	97
Thermal power production (kW)	165	90	0
Fuel consumption mGT/mHAT (kW)	340	340	322
Fuel consumption CC (kW)	98	0	46
Electrical efficiency (%)	21.9	28.2	26.4
Thermal efficiency (%)	37.7	26.6	0
Total efficiency (%)	59.6	54.8	26.4

5. Overview of Available CCUS Technologies

Several storage options are being considered for carbon capture applications, with varying levels of development. Injecting CO_2 into geologic formations could be considered as it is a mature technology already in use in the oil and gas industry. The main concern associated with geological storage is leakage of the concentrated CO_2 stream and associated environmental damage. However, annual leakage rates in the literature range from 0.00001% to 1%. The three main types of geological formation for carbon storage are depleted oil and gas reserves, deep saline reservoirs, and unmineable coal seams [10,11]. The captured CO_2 can also be sold for profit, which is one avenue to be explored in the interest of offsetting the high costs of CCS. Examples of utilization are direct utilization, enhanced oil recovery (EOR), carbonation, and conversion into chemicals and fuels. Direct uses include the food and drink industry, such as for drink carbonation, the decaffeination process, etc. CO_2 can also be used directly for pharmaceutical applications. Direct utilization is only possible for sources with high purity CO_2 waste streams, such as ammonia production. EOR is the process of using CO_2 to extract crude oil from an oil field or natural gas from unmineable coal deposits. As applied to natural gas, this process is still under development and not yet commercially available. However, EOR for crude oil extraction has been used for many years already in both Canada and the USA. For EOR, CO_2 is injected into otherwise unrecoverable oil reserves for increased oil extraction. Most of the CO_2 is pumped back to the surface and recycled, although some is released into the atmosphere. Under certain conditions, the CO_2 injected for EOR could remain underground as in geological storage. As mentioned, captured CO_2 can also be transformed into fuels and useful chemicals, or used as feedstock for fuel production. Unfortunately, using CO_2 for feedstock results in a highly energy intensive process. Further, fuels and chemicals have a short lifespan, and are quickly released back into the atmosphere. Therefore, the benefits of capture are undermined [28,35,42].

For perspective, the following section will provide a brief overview of the large-scale CC projects currently in operation. CCS technology functioning at large-scale has existed since the 1970s. Government subsidized projects to aid develop and commercialize CC plants have been principally concentrated on electricity generation. The major part of CO_2

injection from CC is in EOR in the USA that offers extra revenue to partially offset capture costs. In Canada, there are three capture plants in operation and two in construction as of 2019. The Boundary Dam Carbon Capture and Storage facility in Saskatchewan has been in operation since 2014 with a capture capacity of one million tons of CO_2 on an annual basis [24,101]. The post-combustion capture facility is coupled to a fossil-fueled electricity generating plant, to which CO_2 is largely transported by pipeline and used for EOR at the Weyburn Oil Unit. The remaining CO_2 is transported, also via pipeline, to the nearby Aquistore project for geological storage [22]. The Petra Nova plant in Texas is another commercial large-scale fossil-fueled power plant integrated with CCS technology, with annual CO_2 capture of one million tons. Both plants sell captured CO_2 for EOR, which partially offsets the cost introduced by CCS [107–109].

The micro cogeneration systems with integrated CC discussed thus far are purely in the development stage and are not yet commercially available. For instance, Clean O_2 Carbon Capture has developed the first commercial unit for decentralized CC applications [110]. The product provides direct capture to by-product utilization with minimal processing. The process is described as follows: A portion of the flue gases, having a CO_2 concentration of 40,000 ppm or greater, pass through the reaction chamber where caustic soda reacts with the carbon dioxide to create soda ash and water. The flue gases then pass through the reaction chamber into the heat exchanger where waste heat is recovered by heating the municipal water supply for domestic hot water. The caustic soda must be replenished weekly. The Clean O_2 Carbon Capture unit is currently installed in concentrated residential, commercial, and single residential applications. Depending on the use, either a residential or commercial unit may be installed. For concentrated residential applications, a commercial pilot is installed at Garrison Woods and Marda Loop in Calgary. For commercial use, a commercial unit is installed at Westjet airlines in Calgary. Both are projected to produce 6.5 tonnes of by-product, the equivalent of 3 metric tonnes of captured CO_2 per year. In terms of single residential use, a residential pilot is installed in Calgary and projected to produce 630 kg of by-product, the equivalent of 320 kg of captured CO_2 per year. The cost per tonne of CO_2 captured is approximately $14.94. However, by selling the soda ash by-product, the cost of capture is actually negative. The advantages of using the Clean O_2 Carbon Capture unit are notable, comprising savings of up to 20% on energy charges per annum, and, reflecting ecological concerns, inhibition of GHG emissions into the atmosphere [110].

6. Cost Analysis

This section presents a brief overview of the cost of carbon capture at large scale, as the literature on the small-scale is scarce. This will be to provide a general sense of the costs of CC, which can provide context for evaluating potential costs at a small scale. Further, an economic evaluation of the impact of CC on the biomass-fired ORC cycle for mCHP applications is presented.

For large scale power generation plants, CC costs are typically defined based on the separation and compression costs at a single facility independent of the costs of transport, storage, or further conversion steps [111–115]. Rubin et al. [35] presented an updated CCS cost estimate and compared it to the CCS costs reported in the 2005 Intergovernmental Panel on Climate Change (IPCC) Special Report on Carbon Dioxide Capture and Storage (SRCCS) [15]. This analysis was based on natural gas combined cycle plants with MEA-based post-combustion CC. The previous research details only separation and compression costs based on the addition of the capture plant only, without consideration of transport and storage. The Robin et al. [35] study included the costs of transport and storage to provide an estimate of CCS costs.

All planned large-scale transport of CO_2 continues with pipelines. However, according to Pieri et al. [111], the use of trucks is likely to be more economical for short distance and small-scale CO_2 transport than branch pipelines. Pipelines are advantageous as they can handle large flowrates but impose an economic challenge in the case of low flowrates. For small quantities, truck tankers and railroad tankers are favorable, and provide lower capital costs. Nonetheless, large scale trunk pipelines would need to be constructed for both large and small-scale capture [111]. The current costs of CO_2 pipeline transport taken from the SRCCS [15], IPCC [15], ZEP [112], and USDOE [113], are summarized in Tables 11 and 12 for onshore and offshore pipelines, respectively [35]. The values for SRCCS are reported in 2002 USD/tCO_2/250 km and the values for all other studies are in 2013 USD/tCO_2/250 km. These results are derived for "normal" terrain; however, given that the cost is highly dependent upon the given terrain, actual costs may be much higher. The costs were adjusted to a common basis of 2013 USD/tCO_2/250 km using the Chemical Engineering Plant Cost Index (CEPCI) escalation factors. From 2002 to 2013 the CEPCI rose more than the US general inflation index, CPI (44% vs. 29%, respectively), showing "real" cost escalation during that time. The Power Capital Cost Index (PCCI, an index particular to the capital cost of non-nuclear power plants) increased further, at 64%. As the study of Rubin et al. [35] involved the cost of power plants with and without CCS, they applied the PCCI to escalate the capital cost of power plants from 2002 to 2013 dollars. To adjust transport and storage costs, they utilized the CEPCI, as these services are usually supplied to power plants by distinct companies largely from the oil and gas industry.

Table 11. Transport costs for onshore pipelines at three diverse capacities (adapted from [35]).

Study	3 MtCO_2/Year	10 MtCO_2/Year	30 MtCO_2/Year
SRCCS (2002) [15]	3.0–5.0	1.5–2.6	0.9–1.5
IPCC (2005) [15]	4.3–7.2	2.2–3.7	1.3–2.2
ZEP (2011) [112]	10.9	3.3	-
USDOE (2014) [113]	4.9	-	1.7

Note: The values for SRCCS values are reported in 2002 USD/tCO_2/250 km and the values for all other studies are in 2013 USD/tCO_2/250 km.

Table 12. Transport costs for offshore pipelines at three diverse capacities (adapted from [35]).

Study	3 MtCO_2/Year	10 MtCO_2/Year	30 MtCO_2/Year
SRCCS (2002) [15]	5.0–6.2	2.4–3.0	1.3–1.7
IPCC (2005) [15]	7.2–8.9	3.4–4.3	1.9–2.4
ZEP (2011) [113]	14.8	4.8	-

Note: The values for SRCCS values are reported in 2002 USD/tCO_2/250 km and the values for all other studies are in 2013 USD/tCO_2/250 km.

Rubin et al. [35] also investigated the cost of storage through a survey of the available literature on the topic. They focused only on storage through injection into geologic formations as this technology is well established and has been practiced for many years in different contexts. However, in the context of storage, more research must be done, and several uncertainties remain. For example, it is unknown how future regulations may impact costs in terms of monitoring and liability, as well as the impact of public acceptance on project economics. As discussed, several types of geologic storage reservoirs are available, with varying costs depending on the reservoir type. Based on this, the various studies examined by Rubin et al. [35] provide cost ranges. In the SRCCS, the cost of geologic storage was reported to range from 0.5 to 8.0 2002 USD/tCO_2, with an added monitoring cost of 0.1–0.3 2002 USD/tCO_2. The typical ranges of onshore storage costs evaluated on a standard basis are reported in Table 13 from four different studies.

Table 13. Onshore storage costs on a standard basis (2013 USD/tCO_2) (adapted from [35]).

	Study	IPCC (2005)	ZEP (2011)	GCCSI (2011)	USDOE (2014)
Onshore storage costs (USD/tCO_2)	Low	1	2	6	7
	High	12	18	13	13

Table 14 displays the levelized cost ranges for amine-based post-combustion capture integrated with a natural gas combined cycle (NGCC) plant based on the literature, as reviewed and reported by Rubin et al. [35]. From the analysis, it is clear that the total cost of CCS can be decreased considerably if CO_2 is sold for EOR as well as geologic storage. However, it is worth pointing out that due to the volatility of oil prices, the value of the EOR credits is quite uncertain.

Table 14. Range of total costs for CC, transport and geological storage (2013 USD) (adapted from [35]).

Cost and Performance Parameters	NGCC with Post-Combustion Capture
Reference plant without CCS: Levelized cost of electricity (USD/MWh)	42–83
Power plants with CCS	
Increased fuel requirement per net MWh (%)	13–18
CO_2 captured (kg/MWh)	360–390
CO_2 avoided (kg/MWh)	310–330
% CO_2 avoided	88–89
Power plant with capture, transport and geological storage	
Levelized cost of electricity (USD/MWh)	63–122
Electricity cost increase for CCS (USD/MWh)	19–47
% increase	28–72
Power plant with capture, transport and geological storage with EOR credits	
Levelized cost of electricity (USD/MWh)	48–112
Electricity cost increase for CCS (USD/MWh)	3–37
% increase	7–56

Carbon capture is the most costly phase of a CCUS supply chain, due to high capture cost from sources, which are diluted in CO_2. This is especially true at a small-scale, as CO_2 content in flue gases are significantly lower than those from large scale energy generation. The values of the cost of CO_2 captured and the cost of CO_2 avoided reported by Rubin et al. [35] would be much lower than those for small-scale generation due to the diluted CO_2. For this application, the greatest challenge to be overcome is making MEA-based post combustion economically feasible through lowering of the high energy requirement for regeneration, solvent degradation and loss, and corrosion issues.

For an ORC-based micro-CHP system, at optimal conditions using R245fa as a working fluid, Zhu et al. [102] performed a thermo-economic analysis to assess the cost effects of CO_2 capture integration, with the results summarized in Table 15. It was found that the total investment (INV_{tot}) increases from 31.48 to 82.9 × 10^4 USD and the annual operation and maintenance costs ($C_{O\&M}$) increase from 39.26 to 62.36 × 10^4 USD. Further the net annual income (NAI) and net present value (NPV) decrease, while the dynamic payback period (DPP), net power index (NPI), and levelized energy cost (LEC) increase, according to Table 15. This leads to the conclusion that the economic performance is decreased with MEA-based CC. It is important to mention that the biomass fuel cost represents a substantial portion of the annual operation and maintenance costs, followed by the cost of input heat for MEA absorbent regeneration.

Following the previous study by Zhu et al. [102], a comparison of the equipment cost ratio for a biomass-fired ORC-CHP system with and without CO_2 capture investment, showed the following differences: evaporator (19.65%), condenser (18.14%), expander (16.04%), and biomass boiler (15.79%). Further, the cooling water loop, which consists of the cooling water pump (7.33%) and cooling tower (4.26%), accounted for 11.59% of the total system investment. When the CO_2 capture system was integrated, it accounted for 62.03% of the total investment. Further, the comparison of the combined ratio for biomass-fired ORC-CHP system annual operation and maintenance costs ($C_{O\&M}$) were also provided without and with CO_2 capture, showing that biomass-fuel (CA_{bf}) accounted for

58.07% of the annual operation and maintenance costs. This was followed by the thermal heat expenditures for MEA solvent regeneration in the stripper (27.84%), which resulted from the significant energy penalty.

Table 15. Thermo-economic comparison of biomass-fired ORC-CHP system (R245fa) with and without CC (adapted from [102]).

Parameter	Biomass-Fired ORC-CHP	
	Without CC	With CC
Total investment cost, INV_{tot} ($\times 10^4$ USD)	31.48	82.9
Annual operation and maintenance costs, $C_{O\&M}$ ($\times 10^4$ USD)	39.26	62.36
Net annual income, NAI ($\times 10^4$ USD/year)	159.15	136.05
Dynamic payback period, DPP (year)	0.21	0.62
Profit ratio of investment, PRI (-)	5.06	1.69
Net present value, NPV ($\times 10^4$ USD/year)	1951.88	1663.98
Net power index, NPI (USD/kWe)	1240.73	3267.31
levelized energy cost, LEC (USD/kWh)	0.19	0.31

7. Other Applications of CCS

With the goal of net zero emissions, negative emissions technology (NET) is required for the purpose of recapturing greenhouse gases emitted in the past. Direct air capture is the process of removing CO_2 from the air and generating a concentrated stream of carbon dioxide for sequestration or utilization [116–119]. If the carbon dioxide is stored, DAC contributes to the group of negative emission technologies. A challenge that must be addressed during the development of DAC systems is the large energy input that is required to remove and concentrate CO_2 to a pure stream (>90%) from air (390 ppm). The thermodynamic minimum energy required is 250 kWh per ton of CO_2, which is significantly higher than that for concentrated sources, such as flue gases from power plants. The current cost per ton of CO_2 is estimated to be \$200–\$1000, whereas the cost estimate for concentrated point sources is within the range of \$40–\$60 [108].

Several DAC technologies are currently in developmental stages; however, there is no agreement on the processes of CO_2 separation from air. Several processes are under development as reported in the literature—these include solid sorbent adsorption, vacuum swing adsorption, and calcium (carbonate-bicarbonate) looping absorption cycles [120–126].

Due to the highly dilute concentration of CO_2 in ambient air, the energy input is far greater than that for point source CCUS. However, net negative global emissions are necessary to accomplish the current climate change prevention goals [4,108].

An interesting application for DAC is a ventilation approach for buildings using CO_2 capture. Kim et al. [120,121] proposed a CO_2 adsorption capture device and its use as a ventilation strategy in buildings in 2015, and further examined the system moisture performance in 2020. The goal was to provide a more efficient means of indoor air recirculation. Through numerical calculations, they found that during the tropical summer and central European winter seasons, 30–60% of air ventilation energy for cooling and heating can be saved when compared to the traditional air ventilation system and can save building energy at peak energy load times.

Baus and Nehr [126] proposed a coupling of HVAC-systems with DAC-technology to separate CO_2 in the exhaust air of buildings and recirculate the CO_2-depleted air back into the building. Based on a theoretical method, the possibilities and drawbacks of the recommended HVAC/DAC-coupling in recirculation mode were evaluated. They concluded that the system can decrease the energy demand of buildings while also enabling access to unutilized CO_2-resources transported in the building. In addition, the system presents the possibility of enhancing indoor air quality. However, the authors pointed out that an adequate DAC unit for operation in indoor air is not yet commercially available.

8. Conclusions

This paper has reviewed the latest advances related to small-scale carbon capture systems and their application, focusing mainly on micro-combined heat and power co-generation systems for use in buildings. As can be seen, the use of CC on small-scale distributed combined heat and power sources has the potential to reduce carbon emissions and achieve the carbon neutral goal. This review discussed in detail some of the main concerns and proposed several key findings that can aid in assessing the feasibility of CC when applied to the small-scale.

The findings pertaining to the application of carbon capture in different energy systems in summary are as follows: For micro gas turbine systems, EGR can increase CO_2 concentration by 2.2 times the baseline mGT cycle, and can be increased to a ratio of 0.55, significantly decreasing the volumetric flow rate to the capture plant. This leads to decreased capital and operating costs of the capture plant. S-EGR can enhance CO_2 concentrations further than EGR while maintaining stable combustion. Conversion into the mHAT cycle can increase global efficiency to 25.8% compared to the baseline efficiency of 23.0%. Improved energy integration using turbine exhaust heat for reboiler duty can increase the cycle efficiency up to 26.8% when there is no heat demand. For the hybrid solid state fuel cell and micro gas turbine systems, the maximum net efficiency was found to be 51.4% LHV and a 100% carbon capture ratio can be achieved. The biomass-fired organic Rankine cycle systems allow for negative carbon emissions when integrated with CC. The maximum system efficiency was found to be 13.7%, not including the effects of CC. This must be further assessed. When the CO_2 capture system is integrated, it accounts for 62.0% of the total system investment; thermal expenditures for MEA solvent regeneration account for 27.8% of the annual operating and maintenance costs. Moreover, direct air capture was also identified as a potential technology for negative carbon emissions. However, further development is required to reduce the substantial energy penalty and increase the economic feasibility.

Currently, carbon capture, as applied at small-scale in buildings applications, is in the developmental stages, and further research must be performed to reduce the energy penalty of the capture process for economic feasibility. Additionally, the majority of the results from the literature were obtained through numerical analyses. Therefore, it is necessary that more extensive experimental work be accomplished to validate the current findings. Further, there is a gap in the literature pertaining to the economic feasibility of small-scale carbon capture systems, and, in the future, a full cost analysis must be performed to address this.

Although large-scale CC/CCS/CCUS systems are currently deployed in a range of industrial applications, the integration of small-scale CC in micro-CHP systems for buildings in full-scale projects is required to advance the understanding and practice necessary for demonstration and deployment of this technology. To achieve this, as mentioned above, more research and development are required to develop innovative concepts that have the capacity to improve the operability, reliability and environmental performance, and to substantially decrease the costs of CC for use in new and existing residential, commercial and institutional buildings. Experimental pilot and demonstration scales are required to address these gaps.

The effective deployment of integrated small-scale CC in building energy systems requires the active involvement and engagement of government, building constructers, power utilities, CC manufacturers, policy makers and owners. It is anticipated that this review paper will be of interest to all concerned parties to enhance awareness of the challenges and issues, and to contribute more to this area.

Author Contributions: Conceptualization, W.Y.; methodology, W.Y.; formal analysis, W.Y., E.E. and M.L.; investigation, W.Y., E.E. and M.L.; data curation, E.E. and M.L.; writing—original draft preparation, W.Y., E.E. and M.L.; writing—review and editing, W.Y., E.E. and M.L. All authors have read and agreed to the published version of the manuscript.

Funding: Funding for this research was provided by Natural Resources Canada through the Program of Energy Research and Development.

Institutional Review Board Statement: Not applicable.

Informed Consent Statement: Not applicable.

Data Availability Statement: Data are contained within this review article.

Acknowledgments: The authors would like to thank the Office of Energy Research and Development (OERD) of Natural Resources Canada for their valuable financial support.

Conflicts of Interest: The authors declare no conflict of interest.

Nomenclature

CC	carbon capture
CCS	carbon capture and storage
CCUS	carbon capture, utilization and storage
CH_4	methane
C_2H_6	ethane
CHP	combined heat and power
CO	carbon monoxide
CO_2	carbon dioxide
DAC	direct atmospheric capture
DPP	dynamic payback period
EGR	exhaust gas recirculation
EOR	enhanced oil recovery
FC	fuel cell
GHG	greenhouse gas
GT	gas turbine
HRSG	heat recovery steam generator
HVAC	heating, ventilation, and air conditioning
H_2	hydrogen
IEA	International Energy Agency
IPCC	Intergovernmental Panel on Climate Change
LEC	levelized energy cost
LHV	lower heating value
MEA	monoethanolamine
micro-CHP	micro combined heat and power
mGT	micro gas turbine
mHAT	micro humidified air turbine
NET	negative emissions technology
NGCC	natural gas combined cycle
NPI	net power index
NPV	net present value
NO_x	oxides of nitrogen
O_2	oxygen
ORC	organic Rankine cycle
PACT	pilot-scale advanced capture technology
REVAP	regenerative evaporation
S-EGR	selective exhaust gas recirculation
SOA	state-of-the-art
SOFC	solid oxide fuel cell
SRCCS	Special report on carbon dioxide capture and storage
STIG	steam injected gas turbine
TRCL	three-reactors chemical looping
UHC	unburned hydrocarbon
USDOE	US department of energy
VUB	Vrije Universiteit Brussel
WAC	water atomizing inlet air cooling
ZEP	Zero Emissions Platform

References

1. IEA. Net Zero by 2050, A Roadmap for the Global Energy Sector. 2021. Available online: https://www.iea.org/reports/net-zero-by-2050 (accessed on 15 February 2022).
2. IEA. *World Energy Outlook 2021*; IEA: Paris, France, 2021. Available online: https://www.iea.org/reports/world-energy-outlook-2021 (accessed on 15 February 2022).
3. IRENA. *Global Energy Transformation: A Roadmap to 2050*; International Renewable Energy Agency: Abu Dhabi, United Arab Emirates, 2018; Available online: www.irena.org/publications (accessed on 15 February 2022).
4. Haszeldine, R.S.; Flude, S.; Johnson, G.; Scott, V. *Negative Emissions Technologies and Carbon Capture and Storage to Achieve the Paris Agreement Commitments*; Philosophical Transactions the Royal Society Publishing: Edinburgh, UK, 2018.
5. IEA. Energy Technology Perspectives 2020, Special Report on Carbon Capture, Utilisation and Storage, CCUS in Clean Energy Transitions. 2020. Available online: https://webstore.iea.org/download/direct/4191 (accessed on 15 February 2022).
6. IEA. About CCUS. 2021. Available online: https://www.iea.org/reports/about-ccus (accessed on 15 February 2022).
7. Global CCS Institute. *Global Status of CCS Report*; Fluid Branding: Melbourne, Australia, 2021.
8. Martin-Roberts, V.; Scott, S.; Flude, G.; Johnson, R.S.; Haszeldine, S. Gilfillan, Carbon capture and storage at the end of a lost decade. *One Earth* **2021**, *4*, 1569–1584. [CrossRef]
9. Singh, B.; Strømman, A.H.; Hertwich, E.G. Comparative life cycle environmental assessment of CCS technologies. *Int. J. Greenh. Gas Control* **2011**, *5*, 911–921. [CrossRef]
10. Cuellar-Franca, R.M.; Azapagic, A. Carbon capture, storage and utilisation technologies: A critical analysis and comparison of their life cycle environmental impacts. *J. CO2 Util.* **2015**, *9*, 82–102. [CrossRef]
11. Leung, D.Y.C.; Caramanna, G.; Maroto-Valer, M.M. An overview of current status of carbon dioxide capture and storage technologies. *Renew. Sustain. Energy Rev.* **2014**, *39*, 426–443. [CrossRef]
12. Abanades, J.C.; Arias, B.; Lyngfelt, A.; Mattisson, T.; Wiley, D.E.; Li, H.; Ho, M.T.; Mangano, E.; Brandani, S. Emerging CO_2 capture systems. *Int. J. Greenh. Gas Control* **2015**, *40*, 126–166. [CrossRef]
13. Bui, M.; Adjiman, C.S.; Bardow, A.; Anthony, E.J.; Boston, A.; Brown, S.; Fennel, P.S.; Fuss, S.; Galindo, A.; Hackett, L.A.; et al. Carbon capture and storage (CCS): The way forward. *Energy Environ. Sci.* **2018**, *11*, 1062–1176. [CrossRef]
14. Wang, M.; Oko, E. Special issue on carbon capture in the context of carbon capture, utilisation and storage (CCUS). *Int. J. Coal Sci. Technol.* **2017**, *4*, 1–4. [CrossRef]
15. IPCC (Intergovernmental Panel on Climate Change). *IPCC Special Report on Carbon Dioxide Capture and Storage*; Metz, B., Davidson, O., de Coninck, H.C., Loos, M., Meyer, L.A., Eds.; Prepared by working group III of the intergovernmental panel on climate change; Cambridge University Press: Cambridge, UK; New York, NY, USA, 2005; ISBN 13 978-0-521-86643-9.
16. Wang, M.; Lawal, A.; Stephenson, P.; Sidders, J.; Ramshaw, C. Postcombustion CO_2 capture with chemical absorption: A state-of the-art review. *Chem. Eng. Res. Des.* **2011**, *89*, 1609–1624. [CrossRef]
17. Baena-Moreno, F.M.; Rodríguez-Galán, M.; Vega, F.; Alonso-Fariñas, B.; Arenas, L.F.V.; Navarrete, B. Carbon capture and utilization technologies: A literature review and recent advances. *Energy Sources Part A Recovery Util. Environ. Eff.* **2019**, *41*, 1403–1433. [CrossRef]
18. Jiang, K.; Ashworth, P.; Zhang, S.; Liang, X.; Sun, Y.; Angus, D. China's carbon capture, utilization and storage (CCUS) policy: A critical review 2019. *Renew. Sustain. Energy Rev.* **2019**, *119*, 109601. [CrossRef]
19. Gür, T.M. Carbon dioxide emissions, capture, storage and utilization: Review of materials, processes and technologies. *Prog. Energy Combust. Sci.* **2022**, *89*, 100965. [CrossRef]
20. Hong, W.Y. A techno-economic review on carbon capture, utilisation and storage systems for achieving a net-zero CO2 emissions future. *Carbon Capture Sci. Technol.* **2022**, *3*, 100044. [CrossRef]
21. Hasan, M.M.F.; First, E.L.; Boukouvala, F.; Floudas, C.A. A multi-scale framework for CO_2 capture, utilization, and sequestration: CCUS and CCU. *Comput. Chem. Eng.* **2015**, *81*, 2–21. [CrossRef]
22. Yan, J.; Zhang, Z. Carbon Capture, Utilization and Storage (CCUS). *Appl. Energy* **2019**, *235*, 1289–1299. [CrossRef]
23. Dowell1, N.M.; Fennell, P.S.; Shah, N.; Maitland, G.C. The role of CO_2 capture and utilization in mitigating climate change. *Nat. Clim. Chang.* **2017**, *7*, 243–249. [CrossRef]
24. Peridas, G.; Mordick Schmidt, B. The role of carbon capture and storage in the race to carbon neutrality. *Electr. J.* **2021**, *34*, 106996. [CrossRef]
25. Raza, A.; Gholami, R.; Rezaee, R.; Rasouli, V.; Rabiei, M. Significant aspects of carbon capture and storage—A review. *Petroleum* **2019**, *5*, 335–340. [CrossRef]
26. Karimi, F.; Khalilpour, R. Evolution of carbon capture and storage research: Trends of international collaborations and knowledge maps. *Int. J. Greenh. Gas Control* **2015**, *37*, 362–376. [CrossRef]
27. Vögele, S.; Rübbelke, D.; Mayer, P.; Kuckshinrichs, W. Germany's "No" to carbon capture and storage: Just a question of lacking acceptance? *Appl. Energy* **2018**, *214*, 205–218. [CrossRef]
28. Beck, L. Carbon capture and storage in the USA: The role of US innovation leadership in climate-technology commercialization. *Clean Energy* **2020**, *4*, 2–11. [CrossRef]
29. Zhang, K.; Xie, J.; Li, C.; Hu, L.; Wu, X.; Wang, Y. A full chain CCS demonstration project in northeast Ordos Basin, China: Operational experience and challenges. *Int. J. Greenh. Gas Control* **2016**, *50*, 218–230. [CrossRef]

30. Li, J.; Hou, Y.; Wang, P.; Yang, B. A Review of carbon capture and storage project investment and operational decision-making based on bibliometrics. *Energies* **2019**, *12*, 23. [CrossRef]
31. IEA. 20 Years of Carbon Capture and Storage—Accelerating Future Deployment, International Energy Agency, Paris, France. Available online: https://www.iea.org/publications/freepublications/publication/20-years-ofcarbon-capture-and-storage.html (accessed on 20 February 2022).
32. Quale, S.; Rohling, V. The European Carbon dioxide Capture and Storage Laboratory Infrastructure (ECCSEL). *Green Energy Environ.* **2016**, *1*, 180–194. [CrossRef]
33. MIT. CCS Project Database, Massachusetts Institute of Technology, Boston, USA. Available online: https://sequestration.mit.edu/tools/projects/index_capture.html (accessed on 18 February 2022).
34. NETL. NETL's Carbon Capture and Storage (CCS) Database—Version 5, National Energy Technology Laboratory, USA. Available online: https://www.netl.doe.gov/research/coal/carbon-storage/strategic-program-support/database (accessed on 20 February 2022).
35. Rubin, E.; Davison, J.E.; Herzog, H.J. The cost of CO_2 capture and storage. *Int. J. Greenh. Gas Control* **2015**, *40*, 378–400. [CrossRef]
36. Leeson, D.; Dowell, N.M.; Shah, N.; Petit, C.; Fennell, P.S. A techno-economic analysis and systematic review of carbon capture and storage (CCS) applied to the iron and steel, cement, oil refining and pulp and paper industries, as well as other high purity sources. *Int. J. Greenh. Gas Control* **2017**, *61*, 71–84. [CrossRef]
37. Plaza, M.G.; Martínez, S.; Rubiera, F. CO_2 capture, use, and storage in the cement industry: State of the art and expectations. *Energies* **2020**, *13*, 5692. [CrossRef]
38. Petrakopoulou, F.; Tsatsaronis, G. Can carbon dioxide capture and storage from power plants reduce the environmental impact of electricity generation? *Energy Fuels* **2014**, *28*, 5327–5338. [CrossRef]
39. Wilberforce, T.; Baroutaji, A.; Soudan, B.; Al-Alami, A.H.; Olabi, A.G. Outlook of carbon capture technology and challenges. *Sci. Total Environ.* **2019**, *657*, 56–72. [CrossRef]
40. Pan, S.Y.; Chiang, P.C.; Pan, W.; Kim, H. Advances in state-of-art valorization technologies for captured CO_2 toward sustainable carbon cycle. *Crit. Rev. Environ. Sci. Technol.* **2018**, *48*, 471–534. [CrossRef]
41. Adamsli, T.A., II; Hoseinzade, L.; Madabhushi, P.B.; Okeke, I.J. Comparison of CO_2 capture approaches for fossil-based power generation: Review and meta-study. *Processes* **2017**, *5*, 44. [CrossRef]
42. González-Salazar, M.A. Recent developments in carbon dioxide capture technologies for gas turbine power generation. *Int. J. Greenh. Gas Control* **2015**, *34*, 106–116. [CrossRef]
43. Hetti, R.K.; Karunathilake, H.; Chhipi-Shrestha, G.; Sadiq, R.; Hewage, K. Prospects of integrating carbon capturing into community scale energy systems. *Renew. Sustain. Energy Rev.* **2020**, *133*, 110193. [CrossRef]
44. Liyanage, D.R.; Hewage, K.; Karunathilake, H.; Chhipi-Shrestha, G.; Sadiq, R. Carbon Capture Systems for Building-Level Heating Systems—A Socio-Economic and Environmental Evaluation. *Sustainability* **2021**, *13*, 10681. [CrossRef]
45. Kanniche, M.; Gros-Bonnivard, R.; Jaud, P.; Valle-Marcos, J.; Amann, J.M.; Bouallou, C. Pre-combustion, post-combustion and oxy-combustion in thermal power plant for CO_2 capture. *Appl. Therm. Eng.* **2009**, *30*, 53–62. [CrossRef]
46. Hossain, M.M.; de Lasa, H.I. Chemical-looping combustion (CLC) for inherent CO_2 separation—A review. *Chem. Eng. Sci.* **2008**, *63*, 4433–4451. [CrossRef]
47. Lawal, A.; Wang, M.; Stephenson, P.; Yeung, H. Dynamic modelling of CO_2 absorption for post combustion capture in coal-fired power plants. *Fuel* **2009**, *88*, 2455–2462. [CrossRef]
48. Cousins, A.; Wardhaugh, L.T.; Feron, P.H.M. A survey of process flow sheet modifications for energy efficient CO_2 capture from flue gases using chemical absorption. *Int. J. Greenh. Gas Control* **2011**, *5*, 605–619. [CrossRef]
49. Vega, F.; Baena-Moreno, F.M.; Fernández, L.M.G.; Portillo, E.; Navarrete, B.; Zhang, Z. Current status of CO_2 chemical absorption research applied to CCS: Towards full deployment at industrial scale. *Appl. Energy* **2020**, *260*, 114313. [CrossRef]
50. Asif, M.; Suleman, M.; Haq, I.; Jamal, S.A. Post-combustion CO_2 capture with chemical absorption and hybrid system: Current status and challenges. *Greenh. Gases Sci. Technol.* **2018**, *8*, 998–1031. [CrossRef]
51. Sreedhar, I.; Nahar, T.; Venugopal, A.; Srinivas, B. Carbon capture by absorption—Path covered and ahead. *Renew. Sustain. Energy Rev.* **2017**, *76*, 1080–1107. [CrossRef]
52. Ben-Mansour, R.; Habib, M.A.; Bamidele, O.E.; Basha, M.; Qasem, N.A.A.; Peedikakkal, A.; Laoui, T.; Ali, M. Carbon capture by physical adsorption: Materials, experimental investigations and numerical modeling and simulations—A review. *Appl. Energy* **2016**, *161*, 225–255. [CrossRef]
53. Belmabkhout, Y.; Guillerm, V.; Eddaoudi, M. Low concentration CO_2 capture using physical adsorbents: Are metal–organic frameworks becoming the new benchmark materials? *Chem. Eng. J.* **2016**, *296*, 386–397.
54. Jiang, L.; Gonzalez-Diaz, A.; Ling-Chin, J.; Roskilly, A.P.; Smallbone, A.J. Post-combustion CO_2 capture from a natural gas combined cycle power plant using activated carbon adsorption. *Appl. Energy* **2019**, *245*, 1–15. [CrossRef]
55. Khalilpour, R.; Mumford, K.; Zhai, H.; Abbas, A.; Stevens, G.; Rubin, E.S. Membrane-based carbon capture from flue gas: A review. *J. Clean. Prod.* **2015**, *103*, 286–300. [CrossRef]
56. Zhao, S.; Feron, P.H.M.; Deng, L.; Favre, E.; Chabanon, E.; Yan, S.; Hou, J.; Chen, V.; Qi, H. Status and progress of membrane contactors in post-combustion carbon capture: A state-of-the-art review of new developments. *J. Memb. Sci.* **2016**, *511*, 180–206. [CrossRef]

57. Sreedhar, I.; Vaidhiswaran, R.; Kamani, B.M.; Venugopal, A. Process and engineering trends in membrane based carbon capture. *Renew. Sustain. Energy Rev.* **2017**, *68*, 659–684. [CrossRef]
58. Olajire, A.A. CO_2 capture and separation technologies for end-of-pipe applications—A review. *Energy* **2010**, *35*, 2610–2628.
59. Somehsaraei, H.N.; Majoumerd, M.M.; Breuhaus, P.; Assadi, M. Performance analysis of a biogas-fueled micro gas turbine using a validated thermodynamic model. *Appl. Therm. Eng.* **2014**, *66*, 181–190. [CrossRef]
60. De Paepe, W.; Contino, F.; Delattin, F.; Bram, S.; de Ruyck, J. Optimal waste heat recovery in micro gas turbine cycles through liquid water injection. *Appl. Therm. Eng.* **2014**, *70*, 846–856. [CrossRef]
61. Stathopoulos, P.; Paschereit, C.O. Retrofitting micro gas turbines for wet operation, A way to increase operational flexibility in distributed CHP plants. *Appl. Energy* **2015**, *154*, 438–446. [CrossRef]
62. Ebrahimi, M.; Soleimanpour, M. Design and evaluation of combined cooling, heating and power using micro gas turbine, adsorption chiller and a thermal damping tank in micro scale. *Appl. Therm. Eng.* **2017**, *127*, 1063–1076. [CrossRef]
63. Rist, J.F.; Dias, M.F.; Palman, M.; Zelazo, D.; Cukurel, B. Economic dispatch of a single micro-gas turbine under CHP operation. *Appl. Energy* **2017**, *200*, 1–18. [CrossRef]
64. De Paepe, W.; Montero Carrero, M.; Bram, S.; Parente, A.; Contino, F. Toward Higher Micro Gas Turbine Efficiency and Flexibility—Humidified Micro Gas Turbines: A Review. *ASME J. Eng. Gas Turbines Power* **2018**, *140*, 081702. [CrossRef]
65. Turbec. T100 Microturbine System: User manual, Technical Description—T100 Natural Gas. D14127–03. Version 3, 09/12/29. 2009. Available online: https://manualzz.com/doc/33686173/t100-microturbine-system-technical-description-t100-natur. (accessed on 18 February 2022).
66. Giorgetti, S.; Bricteux, L.; Parente, A.; Blondeau, J.; Contino, F.; de Paepe, W. Carbon capture on micro gas turbine cycles: Assessment of the performance on dry and wet operations. *Appl. Energy* **2017**, *207*, 243–253. [CrossRef]
67. Giorgetti, S.; de Paepe, W.; Bricteux, L.; Parente, A.; Contino, F. Carbon capture on a micro gas turbine: Assessment of the performance. In Proceedings of the 8th International Conference on Applied Energy—ICAE2016, Bejing, China, 10 August 2016.
68. De Paepe, W.; Carrero, M.M.; Giorgetti, S.; Parente, A.; Bram, S.; Contino, F. Exhaust gas recirculation on humidified flexible micro gas turbines for carbon capture applications. In Proceedings of the ASME Turbo Expo 2016, Seoul, Korea, 13–17 June 2016. no. ASME GT2016-57265.
69. Akram, M.; Ali, U.; Best, T.; Blakey, S.; Finney, K.N.; Pourkashanian, M. Performance evaluation of PACT Pilot-plant for CO_2 capture from gas turbines with Exhaust Gas Recycle. *Int. J. Greenh. Gas Control* **2016**, *47*, 37–150. [CrossRef]
70. Majoumerd, M.M.; Somehsaraei, H.N.; Assadi, M.; Breuhaus, P. Micro gas turbine configurations with carbon capture—Performance assessment using a validated thermodynamic model. *Appl. Therm. Eng.* **2014**, *73*, 172–184. [CrossRef]
71. Ali, U.; Best, T.; Finney, K.N.; Palma, C.F.; Hughes, K.J.; Ingham, D.B.; Pourkashanian, M. Process simulation and thermodynamic analysis of a micro turbine with post-combustion CO_2 capture and exhaust gas recirculation. *Energy Procedia* **2014**, *63*, 986–996. [CrossRef]
72. Ali, U.; Font-Palma, C.; Somehsaraei, H.N.; Majoumerd, M.M.; Akram, M.; Akram, M.; Finney, K.N.; Best, T.; Said, N.B.M.; Assadi, M.; et al. Benchmarking of a micro gas turbine model integrated with post-combustion CO_2 capture. *Energy* **2017**, *126*, 475–487. [CrossRef]
73. Bellas, J.-M.; Finney, K.N.; Diego, M.E.; Ingham, D.; Pourkashanian, M. Experimental investigation of the impacts of selective exhaust gas recirculation on a micro gas turbine. *Int. J. Greenh. Gas Control* **2019**, *90*, 102809. [CrossRef]
74. Darabkhani, H.G.; Jurado, N.; Prpich, G.; Oakey, J.E.; Wagland, S.T.; Anthony, E.J. Design, process simulation and construction of a 100 kW pilot-scale CO_2 membrane rig: Improving in situ CO_2 capture using selective exhaust gas recirculation (S-EGR). *J. Nat. Gas Sci. Eng.* **2018**, *50*, 128–138. [CrossRef]
75. De Paepe, W.; Carrero, M.M.; Bram, S.; Contino, F.; Parente, A. Waste heat recovery optimization in micro gas turbine applications using advanced humidified gas turbine cycle concepts. *Appl. Energy* **2017**, *207*, 218–229. [CrossRef]
76. De Paepe, W.; Carrero, M.M.; Bram, S.; Contino, F. T100 micro gas turbine converted to full humid air operation—A thermodynamic performance analysis. In Proceedings of the ASME Turbo Expo 2015, Montreal, QC, Canada, 19 June 2015. no. ASME GT2015-56673, V003T06A015.
77. Best, T.; Finney, K.N.; Ingham, D.B.; Pourkashanian, M. CO_2-enhanced and humidified operation of a micro-gas turbine for carbon capture. *J. Clean. Prod.* **2018**, *176*, 370–381. [CrossRef]
78. Giorgetti, S.; Parente, A.; Bricteux, L.; Contino, F.; de Paepe, W. Carbon Clean Combined Heat and Power Production from micro Gas Turbines: Thermodynamic Analysis of Different Scenarios. *Energy Procedia* **2017**, *142*, 1622–1628. [CrossRef]
79. Giorgetti, S.; Parente, A.; Bricteux, L.; Contino, F.; de Paepe, W. Optimal design and operating strategy of a carbon-clean micro gas turbine for combined heat and power applications. *Int. J. Greenh. Gas Control* **2019**, *88*, 469–481. [CrossRef]
80. Sammes, N.M.; Boersma, R. Small-scale fuel cells for residential applications. *J. Power Sources* **2000**, *86*, 98–110. [CrossRef]
81. Kazempoor, P.; Dorer, V.; Weber, A. Modelling and evaluation of building integrated SOFC systems. *Int. J. Hydrog. Energy* **2011**, *36*, 13241–13249. [CrossRef]
82. Allane, K.; Saari, A.; Ugursal, I.; Good, J. The financial viability of an SOFC cogeneration system in single-family dwellings. *J. Power Sources* **2006**, *158*, 403–416. [CrossRef]
83. Staffell, I.; Green, R. The cost of domestic fuel cell micro-CHP systems. *Int. J. Hydrog. Energy* **2013**, *38*, 1088–10102. [CrossRef]
84. Haseli, Y.; Dincer, I.; Naterer, G. Thermodynamic modeling of a gas turbine cycle combined with a solid oxide fuel cell. *Int. J. Hydrog. Energy* **2008**, *33*, 5811–5822. [CrossRef]

85. Mehrpooya, M.; Akbarpour, S.; Vatani, A.; Rosen, M.A. Modeling and optimum design of hybrid solid oxide fuel cell-gas turbine power plants. *Int. J. Hydrog. Energy* **2014**, *39*, 21196–21214. [CrossRef]
86. Cheddie, D.F. Thermo-economic optimization of an indirectly coupled solid oxide fuel cell/gas turbine hybrid power plant. *Int. J. Hydrog. Energy* **2011**, *36*, 1702–1709. [CrossRef]
87. Zabihian, F.; Fung, A.S. Performance analysis of hybrid solid oxide fuel cell and gas turbine cycle: Application of alternative fuels. *Energy Convers. Manag.* **2013**, *76*, 571–580. [CrossRef]
88. Isfahani, S.N.R.; Sedaghat, A. A hybrid micro gas turbine and solid state fuel cell power plant with hydrogen production and CO_2 capture. *Int. J. Hydrog. Energy* **2016**, *41*, 9490–9499. [CrossRef]
89. Chaney, L.J.; Tharp, M.R.; Wolf, T.W.; Fuller, T.A.; Hartvigson, J.J. *Fuel Cell/Micro-Turbine Combined cycle, DOE Contract: DE-AC26-98FT40454, Final Report*; McDermott Technology, Inc.: Alliance, OH, USA; Northern Research and Engineering Corporation: Portsmouth, NH, USA, 1999.
90. Liu, A.; Weng, Y. Performance analysis of a pressurized molten carbonate fuel cell/micro-gas turbine hybrid system. *J. Power Sources* **2010**, *195*, 204–213. [CrossRef]
91. Costamagna, P.; Magistri, L.; Massardo, A.F. Design and part-load performance of a hybrid system based on a solid oxide fuel cell reactor and a micro gas turbine. *J. Power Sources* **2001**, *96*, 352–368. [CrossRef]
92. Rajashekara, K. Hybrid fuel-cell strategies for clean power generation. *IEEE Trans. Ind. Appl.* **2005**, *41*, 682–689. [CrossRef]
93. Basrawi, M.F.B.; Yamada, T.; Nakanishi, K.; Katsumata, H. Analysis of the performances of biogas-fuelled micro gas turbine cogeneration systems (MGT-CGSs) in middle-and small-scale sewage treatment plants: Comparison of performances and optimization of MGTs with various electrical power outputs. *Energy* **2012**, *38*, 291–304. [CrossRef]
94. Kupechi, J. Off-design analysis of a micro-CHP unit with solid oxide fuel cells fed by DME. *Int. J. Hydrog. Energy* **2015**, *40*, 12009–12022. [CrossRef]
95. Chan, C.W.; Ling-Chin, J.; Roskilly, A.P. A review of chemical heat pumps, thermodynamic cycles and thermal energy storage technologies for low grade heat utilisation. *Appl. Therm. Eng.* **2013**, *50*, 1257–1273. [CrossRef]
96. Quoilin, S.; Broek, M.V.D.; Declaye, S.; Dewallef, P.; Lemort, V. Techno-economic survey of Organic Rankine Cycle (ORC) systems. *Renew. Sustain. Energy Rev.* **2013**, *22*, 168–186. [CrossRef]
97. Mahmoudi, A.; Fazli, M.; Morad, M.R. A recent review of waste heat recovery by organic Rankine cycle. *Appl. Therm. Eng.* **2018**, *143*, 660–675. [CrossRef]
98. Tocci, L.; Pal, T.; Pesmazoglou, I.; Franchetti, B. A small scale organic Rankine cycle (ORC): A techno-economic review. *Energies* **2017**, *10*, 413. [CrossRef]
99. Rahbar, K.; Mahmoud, S.; Dadah, R.K.; Moazami, N.; Mirhadizadeh, S.A. Review of organic Rankine cycle for small-scale applications. *Energy Convers. Manag.* **2017**, *134*, 135–155. [CrossRef]
100. Dong, L.L.; Liu, H.; Riffat, S. Development of small-scale and micro-scale biomass fuelled CHP systems—A literature review. *Appl. Therm. Eng.* **2009**, *29*, 2119–22126. [CrossRef]
101. Pereira, J.S.; Ribeiro, J.B.; Mendes, R.; Vaz, G.C.; André, J.C. ORC based micro-cogeneration systems for residential application—A state of the art review and current challenges. *Renew. Sustain. Energy Rev.* **2018**, *92*, 728–743. [CrossRef]
102. Zhu, Y.; Li, W.; Li, J.; Li, H.; Wang, Y.; Li, S. Thermodynamic analysis and economic assessment of biomass-fired organic Rankine cycle combined heat and power system integrated with CO_2 capture. *Energy Convers. Manag.* **2020**, *204*, 112310. [CrossRef]
103. Giorgetti, S.; Coppitters, D.; Contino, F.; de Paepe, W.; Bricteux, L.; Aversano, G.; Parente, A. Surrogate-assisted modeling and robust optimization of a micro gas turbine plant with carbon capture. *J. Eng. Gas Turbines Power* **2020**, *142*, 011010. [CrossRef]
104. Best, T.; Finney, K.N.; Ingham, D.B.; Pourkashanian, M. Impact of CO_2-enriched combustion air on micro-gas turbine performance for carbon capture. *Energy* **2016**, *115*, 1138–1147. [CrossRef]
105. Carrero, M.M.; de Paepe, W.; Magnusson, J.; Parente, A.; Bram, S.; Contino, F. Experimental characterisation of a micro Humid Air Turbine: Assessment of the thermodynamic performance. *Appl. Therm. Eng.* **2017**, *118*, 796–806. [CrossRef]
106. Agbonghae, E.O.; Best, T.; Finney, K.N.; Palma, C.F.; Hughes, K.J.; Pourkashanian, M. Experimental and Process Modelling Study of Integration of a Micro-turbine with an Amine Plant. *Energy Procedia* **2014**, *63*, 1064–1073. [CrossRef]
107. The Global Energy Institute. 2020. Available online: http://status.globalccsinstitute.com/ (accessed on 20 February 2022).
108. SAPEA. *Novel Carbon Capture and Utilisation Technologies*; SAPEA: Berlin, Germany, 2018.
109. Folger, P. *Carbon Capture and Sequestration (CCS) in the United States*; Congressional Research Service: Washington, DC, USA, 2018.
110. CleanO2—Residential and Commercial Carbon Capture Unit. Available online: http://cleano2.ca/ (accessed on 20 February 2022).
111. Pieri, T.; Nikitas, A.; Castillo-Castillo, A.; Angelis-Dimakis, A. Holistic Assessment of Carbon Capture and Utilization Value Chains. *Environments* **2018**, *5*, 108. [CrossRef]
112. ZEP (Zero Emissions Platform). *The Costs of CO2 Transport: Post-Demonstration CCS in the EU*; European Technology Platform for Zero Emission Fossil Fuel Power Plants: Brussels, Belgium, 2011.
113. USDOE. *FE/NETL CO2 Transport Cost Model: Description and User's Manual*; Report No. DOE/NETL-2014/1660; US Dept of Energy, National Energy Technology Laboratory: Pittsburgh, PA, USA, 2014.
114. Kuramochi, T.; Ramirez, A.; Turkenburg, W.; Faaij, A. Techno-economic prospects for CO_2 capture from distributed energy systems. *Renew. Sustain. Energy Rev.* **2013**, *19*, 328–347.
115. Loria, P.; Bright, M.B.H. Lessons captured from 50 years of CCS projects. *Electr. J.* **2021**, *34*, 106998. [CrossRef]

116. Sanz-Perez, E.S.; Murdock, C.R.; Didas, S.A.; Jones, C.W. Direct capture of CO_2 from ambient air. *Chem. Rev.* **2016**, *116*, 11840–11876.
117. Shabbani, H.J.K. A review of CO_2 adsorption from ambient air (direct air capture). *Al-Qadisiyah J. Eng. Sci.* **2020**, *13*, 1–6. [CrossRef]
118. Fasihi, M.; Efimova, O.; Breyer, C. Techno-economic assessment of CO_2 direct air capture plants. *J. Clean. Prod.* **2019**, *224*, 957–980. [CrossRef]
119. Keith, D.W.; Holmes, G.; Angelo, D.S.; Heidel, K. A process for capturing CO_2 from the atmosphere. *Joule* **2018**, *2*, 2179. [CrossRef]
120. Kim, M.K.; Baldini, L.; Leibundgut, H.; Wurzbacher, J.A.; Piatkowski, N. A novel ventilation strategy with CO_2 capture device and energy saving in buildings. *Energy Build.* **2015**, *87*, 134–141. [CrossRef]
121. Kim, M.K.; Baldini, L.; Leibundgut, H.; Wurzbacher, J.A. Evaluation of the humidity performance of a carbon dioxide (CO_2) capture device as a novel ventilation strategy in buildings. *Appl. Energy* **2020**, *259*, 112869. [CrossRef]
122. Gall, E.T.; Nazaroff, W.W. New directions: Potential climate and productivity benefits from CO_2 capture in commercial buildings. *Atmos. Environ.* **2015**, *103*, 378–380. [CrossRef]
123. Viebahn, P.; Scholz, A.; Zelt, O. The potential role of direct air capture in the German energy research program—Results of a multi-dimensional analysis. *Energies* **2019**, *18*, 3443. [CrossRef]
124. Deng, Y.; Li, J.; Miao, Y.; Izikowitz, D. A comparative review of performance of nanomaterials for Direct Air Capture. *Energy Rep.* **2021**, *7*, 3506–3516. [CrossRef]
125. Sabatino, F.; Grimm, A.; Gallucci, F.; van Sint Annaland, M.; Kramer, G.J.; Gazzani, M. A comparative energy and costs assessment and optimization for direct air capture technologies. *Joule* **2021**, *5*, 2047–2076. [CrossRef]
126. Baus, L.; Nehr, S. Potentials and limitations of direct air capturing in the built environment. *Build. Environ.* **2022**, *208*, 108629. [CrossRef]

energies

Communication

CO_2 Capture by Virgin Ivy Plants Growing Up on the External Covers of Houses as a Rapid Complementary Route to Achieve Global GHG Reduction Targets

Jaroslaw Krzywanski [1], Waqar Muhammad Ashraf [2,3], Tomasz Czakiert [4], Marcin Sosnowski [1,*], Karolina Grabowska [1], Anna Zylka [1], Anna Kulakowska [1], Dorian Skrobek [1], Sandra Mistal [1] and Yunfei Gao [5]

1 Faculty of Science and Technology, Jan Dlugosz University in Czestochowa, Armii Krajowej 13/15, 42-200 Czestochowa, Poland; j.krzywanski@ujd.edu.pl (J.K.); k.grabowska@ujd.edu.pl (K.G.); a.zylka@ujd.edu.pl (A.Z.); a.kulakowska@ujd.edu.pl (A.K.); d.skrobek@ujd.edu.pl (D.S.); lenka31121998@interia.pl (S.M.)

2 Department of Mechanical Engineering, University of Engineering and Technology, Lahore 54890, Punjab, Pakistan; engr.waqar986@gmail.com

3 Department of Chemical Engineering, University College London, Gower Street, London WC1E 6BT, UK

4 Department of Advanced Energy Technologies, Faculty of Infrastructure and Environment, Czestochowa University of Technology, Dabrowskiego 73, 42-200 Czestochowa, Poland; tczakiert@is.pcz.czest.pl

5 Department of Chemical and Biomolecular Engineering, North Carolina State University, Raleigh, NC 27607, USA; ygao9@ncsu.edu

* Correspondence: m.sosnowski@ujd.edu.pl

Citation: Krzywanski, J.; Ashraf, W.M.; Czakiert, T.; Sosnowski, M.; Grabowska, K.; Zylka, A.; Kulakowska, A.; Skrobek, D.; Mistal, S.; Gao, Y. CO_2 Capture by Virgin Ivy Plants Growing Up on the External Covers of Houses as a Rapid Complementary Route to Achieve Global GHG Reduction Targets. *Energies* **2022**, *15*, 1683. https://doi.org/10.3390/en15051683

Academic Editors: Marta González Plaza and Rui P. P. L. Ribeiro

Received: 23 January 2022
Accepted: 22 February 2022
Published: 24 February 2022

Abstract: Global CO_2 concentration level in the air is unprecedently high and should be rapidly and significantly reduced to avoid a global climate catastrophe. The work indicates the possibility of quickly lowering the impact of changes that have already happened and those we know will happen, especially in terms of the CO_2 emitted and stored in the atmosphere, by implanting a virgin ivy plant on the available area of walls and roofs of the houses. The proposed concept of reducing CO_2 from the atmosphere is one of the technologies with significant potential for implementation entirely and successfully. For the first time, we showed that the proposed concept allows over 3.5 billion tons of CO_2 to be captured annually directly from the atmosphere, which makes even up 6.9% of global greenhouse gas emissions. The value constitutes enough high CO_2 reduction to consider the concept as one of the applicable technologies allowing to decelerate global warming. Additional advantages of the presented concept are its global nature, it allows for the reduction of CO_2 from all emission sources, regardless of its type and location on earth, and the fact that it will simultaneously lower the air temperature, contribute to oxygen production, and reduce dust in the environment.

Keywords: carbon footprint; carbon capture and storage; zero/low emission building; greenhous effect; environmental protection

1. Introduction

Progressing climate change has nowadays become a fact [1,2]. The carbon dioxide (CO_2) concentration level in the atmosphere is unprecedentedly high and is a crucial driver of catastrophic global warming and climate change calamities. It is required to reduce CO_2 emission under the joint efforts and commitments made by the researchers, governments, and society for humanity's sustainable environment and well-being [3]. The primary factor responsible for global warming is greenhouse gases produced by human activity, mainly carbon dioxide. The above-mentioned activities can be expressed in deforestation for urban living, large-scale industrialization, rapid economic development, and lifestyle [4]. Meeting the energy demands to sustain the growth of human societies is imperative, which is effectively met by the naturally available energy sources. In this regard, it is essential to mention that coal and oil-based energy systems have remained the key technologies for meeting the energy needs of the societies and are still in use across the under-developed and some of the developed nations such as China and the USA [4].

Based on the latest research results, the world emits 51 billion tons of greenhouse gases per year, mainly carbon dioxide, into the atmosphere [5,6]. About 81% of the totally emitted greenhouse gases are carbon dioxide, 10% methane, 7% nitrous oxide [7], and 2% constitute other greenhouse gasses. That is why most scenarios that limit global warming to two degrees Celsius rely on CO_2 removal and storage. Therefore, looking for an effective solution to eliminating CO_2 from the atmosphere is an urgent task, avoiding global warming and climate collapse [8].

In its latest report, the international energy agency (IEA) has presented the levels and sources of CO_2 emissions from various energy resources and industrial sectors [9]. The contribution of various energy resources like renewables, nuclear, natural gas, oil, and gas on the energy production in TWH as well as gigaton of CO_2 (Gt. CO_2) emissions from the electricity production, industrial systems, transport, buildings and other sectors from 2020 to 2050 for the advanced and emerging economies are presented in Figure 1. Increasing penetration of renewables and the associated reduction in CO_2 emissions is projected for the developed economies. At the same time, the demand for coal and gas appears to be nearly constant for energy production for emerging economies. Large volumes of CO_2 emissions are expected to be discharged into the atmosphere.

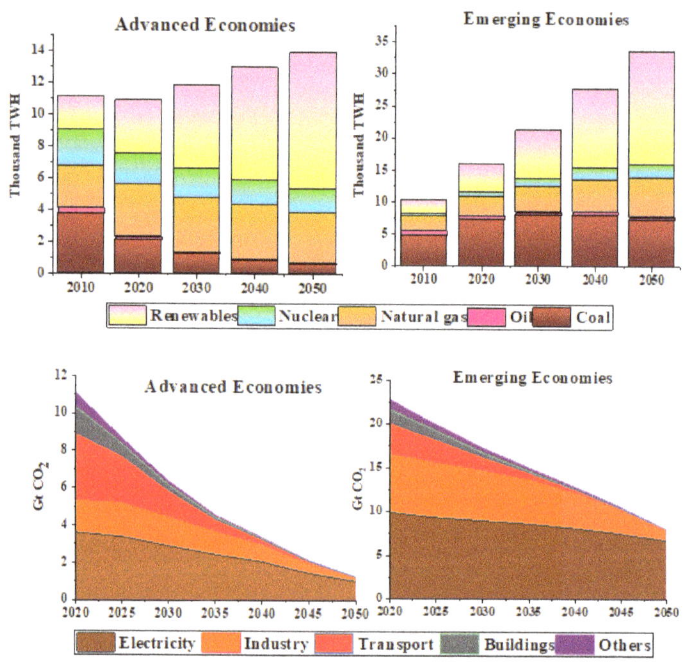

Figure 1. Energy demand and CO_2 emissions trend from the advanced and emerging economies until 2050 [10].

The report highlights that 63.1% of the global energy supply is met by fossil fuels, out of which coal contributes to about 36.7% for power production in 2019. Furthermore, the global energy demand is expected to rise above 80% by 2050. Interestingly, under-developed countries would share 85% of the projected energy needs mainly by the oil and coal-based energy conversion technologies [10]. Therefore, the mitigation of hazardous and greenhouse gas emissions is a significant challenge across the globe. Although, IEA has presented the key technologies such as carbon capture and storage, electrification of the industrial systems, biofuels, etc., as an alternative energy source for decarbonizing the energy sector [9,11]. However, the technologies are under the development stage. In this regard, various global commitments and agreements like Paris Accord and net-zero emissions targets are signed to control the CO_2 emissions from the diverse industrial sectors and seek the timeframe from the nations for reaching the net-zero emissions targets and limiting the global atmospheric temperature rise lower than 1.5 degrees.

Two strategies for reducing atmospheric concentrations of CO_2 can be distinguished. The first one covers most efforts and is focused on reducing emissions of CO_2 to the atmosphere, including increasing energy efficiency or switching to low- or zero-carbon fuel sources. Another approach is to deploy negative emissions technologies (NETs) to remove carbon from the atmosphere and sequester it reliably [8].

The CO_2 capture technologies for energy systems are classified into four groups: pre-combustion capture, oxy-fuel combustion, chemical looping combustion, post-combustion capture [12–16]. Pre-combustion capture is a fuel-reforming system for H_2 production from carbonaceous fuels via gasification and CO_2 separation. In oxy-fuel combustion, N_2 in the air is removed before combustion so that the flue gas consists mainly of CO_2 and H_2O [17]. Chemical looping combustion is based on the reduction and oxidation of metal oxides (oxygen carriers) using fuel (in fuel reactor) and air (in air reactor), respectively, in separate gas streams [18]. Post-combustion capture technologies are gas separation processes to separate CO_2 from flue gas, mainly in air-blown combustors. However, the capacity of all methods is not sufficient, and there is currently no satisfactory method of carbon sequestration [19]. For example, the carbon capture and storage (CCS) technology significantly increases investment costs.

On the other hand, oxy-fuel combustion and chemical looping combustion technologies increase the CO_2 content in flue gas, lowering the nominal cost of CO_2 sequestration. However, they are still suitable only for the CO_2 capture at the gas supply point [19]. Finally, many forms of renewable energy are considered safe, abundant, and clean to use compared to fossil fuels, but location-specific and require storage capabilities [20].

Therefore, this work presents the concept of direct CO_2 capture from the atmosphere through walls and roofs covered with vegetation.

It has to be mentioned, that the only organisms capable of carrying out the photosynthetic process are plants, algae, and a group of bacteria called cyanobacteria [21–23].

The photosynthesis process consists of a light and a dark phase. In the light phase, water is split using light into oxygen, protons and electrons. Then in the dark phase, the protons and electrons are used to reduce CO_2 to carbohydrate. Photosynthesis is a biochemical process that sustains the biosphere as the basis for the food chain. Therefore, plants are the basis of food for most living organisms. Other important roles of plants are the production of oxygen and the reduction of carbon dioxide from the atmosphere [21,24,25].

The amount of carbon sequestrated by plants during photosynthesis is strictly dependent on the area of greenery and plant type [26]. There are many applications of plants known in the literature for the purpose of CO_2 sequestration [27–29]. However, ivy exhibits the best qualities, such as high frost resistance and resistance to air pollution, does not require fertile soil, and grows both in sunny and shaded positions. Its low requirements and high adaptability allow it to occur in various climatic zones [30,31].

The above-mentioned excellent properties of ivy contributed to its selection for the presented analysis concerning CO_2 capture from the atmosphere through walls and roofs covered with ivy plants.

Moreover, it has to be mentioned, that there are many species of climbing plants that can cover walls and roofs of buildings. Each species is characterised by a different strategy of permanent attachment to vertical or horizontal surfaces. In order to exclude the negative influence of plants on the building, it is necessary to analyse the characteristics of individual species.

The attachment pads of Boston ivy (Parthenocissus tricuspidata), the attachment roots of ivy (Hedera helix) and the clustered attachment roots of trumpet creeper (Campsis radicans) can have a negative effect on wall surfaces. However, tendrils tipped with small strongly adhesive pads of five-leaved ivy named as Virginia creeper (Parthenocissus quinquefolia), which rhizomes do not disturb the continuity of the walls, do not have a negative effect on the building structure [32,33].

In addition, ivies belong to plants that rarely suffer from parasitic diseases and are rarely attacked by pests [34]. They can also have a positive impact on the energy efficiency of the building because they provide insulation. During summer they protect against overheating and provide coolness, and in winter they can additionally shield the house from strong winds and frost.

Although the proposed idea belongs to so-called direct-air capture (DAC) technologies, the material's outstanding novelty and broad impact lie in the presented method's designated potential and distinct advantages. The proposed concept further integrates with the global efforts made by the communal populations in the context of mitigation of CO_2 from the atmosphere. The communication of the potential idea to the more significant segments of the population is a crucial step to inclusively contribute to environmental protection with the adoption of green and feasible methods for CO_2 absorption. The obtained results for the first time demonstrate the significant ability of the proposed idea of rapidly globally reducing CO_2 emissions in a relatively shorter time perspective from all emission sources, regardless of its type and location on earth.

2. Methods and Discussion

The proposed concept of removing carbon dioxide directly from the atmosphere through large-scale utilization of green vegetation plants, such as virgin ivy, is shown in Figure 2. It is proposed that the walls and roofs of the houses can be covered with the widely available virgin ivy plants, which would act as the CO_2 capture system. CO_2 in the air is captured by the plant and undergoes the photosynthesis process for converting it into glucose, an energy source for the plants. Moreover, O_2 is released into the air as the by-product of the photosynthesis reaction. Therefore, covering the houses with the virgin ivy plant has several benefits: CO_2 removal from the atmosphere, O_2 release in the air for improving the air quality conditions for breathing purposes and reduction of the hot weather conditions impact on the houses thereby reducing the cooling load in the summer season.

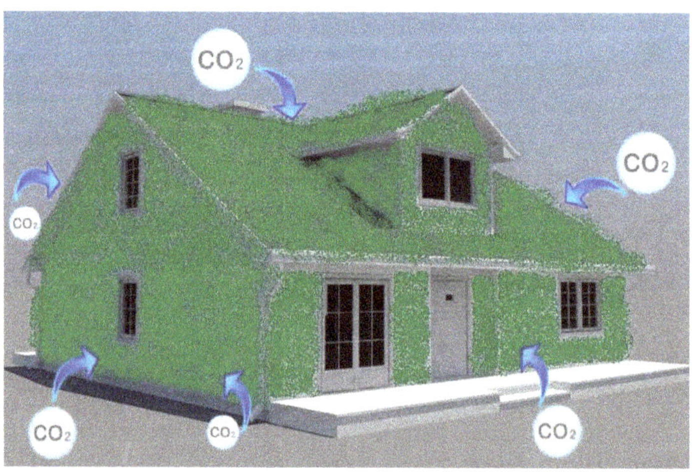

Figure 2. CO_2 absorption concept by virgin ivy plant.

The calculations are based on accessible data reported in the literature. Many parameters are included to estimate the CO_2 capture level from the atmosphere by the virgin ivy plants. Some of the critical parameters mentioned here are the total amount of CO_2 emitted from 1–2 car tank fillings, the average size of oil tanks for cars, the average CO_2 emissions for gasoline and diesel-based engine fuels, the average number of people living in the houses and the world population. The systematic and step-by-step approach incorporating all relevant performance parameters is adopted, and the detailed calculation procedure is shown in Table 1.

Table 1. Calculating procedure of percentage reduction in the CO_2 emissions to the atmosphere.

No.	Parameter	Value		
1.	The average size of oil tanks			
	The average size of fuel tanks for smaller cars (gallons)		12	
	The average size of fuel tanks for larger cars (gallons)		15–16	
	The average size of oil tanks for cars (gallons)		14	
2.	The average CO_2 emissions from a gallon of gasoline			
	The CO_2 emissions from a gallon of gasoline (g/gallon/year)		8887	
	The CO_2 emissions from a gallon of diesel (g/gallon/year)		10,180	
	The average of the CO_2 emissions for two fuels (g/gallon/year)		9533.5	
3.	The CO_2 emissions for the average tank size of 14 gallons		133,469	
4.	Since 1–2 car tank fillings worth of CO_2 will have been absorbed by the plants from the air, per house wall, let us consider three cases with the lowest, average, and maximum number of cars equivalent for CO_2 capture, i.e., 1.0, 1.5 and 2.0	1.0	1.5	2.0
5.	CO_2 emissions absorbed (ton/gallon/year) by a house	0.200193	1.001018	1.334690
6.	Average Number of people living in houses		3	
7.	World Population (billion)		7.9	
8.	The average number of houses (billion)		2.633333	
9.	CO_2 captured (billion ton/year)	0.527175	2.636013	3.514684
10.	Global CO_2 absorbed by ivy plant per year (%)	1.03	5.17	6.89

Because the total amount of CO_2 emitted from 1–2 car tank fillings will have been absorbed by the virgin ivy plants, which have been covered on average-sized plants from the air per average house wall considering the geographical climatic conditions and the phenomena like convective mixing of the atmosphere, wind force and diffusion of CO_2 [18], let us consider three cases with the lowest, average and the maximum number of cars equivalent for CO_2 capture, i.e., 1.0, 1.5 and 2.0. The average size of oil tanks for vehicles equals 14 gallons (as an average of 12 gallons for smaller cars and 15–16 gallons for larger cars) [35]. The average size of the fuel tanks is taken to estimate the vehicles' oil filling capacity. It is found that the average CO_2 emissions for gasoline and diesel-based fuels are 9533.5 g/gallon/year, which is calculated by averaging CO_2 emissions from a gallon of gasoline (8887 g/gallon/year) and diesel (10180 g/gallon/year) [36]. Therefore, the minimum, average and maximum CO_2 capture per house walls and roof can contribute will equal the 0.200193, 1.001018 and 1.334690 ton/gallon/year, respectively. The presented results consider global CO_2 emissions, from all possible sources.

It is quite an overwhelming number for the CO_2 reduction for a single house. Therefore, joint and integrated efforts made by the global human community can significantly synergise the efforts for removing CO_2 from the atmosphere. For this, the CO_2 abatement by the human housing structures in the world covered with virgin ivy plants is also investigated.

Assuming the number of people living in a family in various regions of the world is equal to three [37], and a world population by the end of 2021 is estimated to be equal to 7.9 billion [38], the average number of houses in the world will be equal to 2.633333 billion.

It is further estimated that covering the roofs and walls of all the houses of the world with the virgin ivy plants can mitigate the 0.527175, 2.636013 and 3.514684 billion ton/year for the case corresponding to a minimum, average and maximum of carbon capture, which constitutes about 1.3, 5.2 and 6.89% of global annual carbon dioxide emissions.

Thus, the calculations and analyses showed that the global annual carbon dioxide capture by the proposed concept is much more than the 500 million tons a year, i.e., roughly above 1% of global emissions and above the threshold below which technologies should not compete for the limited resources we have [6]. The general expression for the CO_2 absorption adopted is expressed as:

$$\%\text{ global } CO_2 \text{ absorbed by ivy plant/year} = \frac{\text{Avg. } CO_2 \text{ capacity } \left(\frac{\text{ton}}{\text{year}}\right) * (\text{no. of walls} + \text{roof}) * \text{no. of houses}}{\text{global } CO_2 \text{ load (ton)} * 10000} \quad (1)$$

where the average CO_2 capacity refers to the annually averaged CO_2 absorbed by the virgin ivy plant.

Moreover, it is worth noting that in the case of the minimum CO_2 adsorption capacity, the obtained value of global adsorption exceeds this profitability threshold.

Besides capturing a significant amount of CO_2, the discussed concept also allows lowering the air temperature, contributes to oxygen production and reduces dust in the built-up environment [39].

The significant potential of the virgin ivy plants in removing the CO_2 from the atmosphere can result in the integrated and joint efforts made by the communities around the globe. There is a need to communicate this much potential of the green plants in removing the CO_2 from the air, which directly contributes to the global net-zero emissions targets of the nations to limit the global atmospheric temperature rise to 1.5 degrees. The proposed idea and the advanced carbon capture technologies can collectively work on the single objective to reduce the greenhouse gas emissions level and, therefore, ensure environmental sustainability and the existence of life on the planet.

3. Conclusions

An idea of direct CO_2 capture from the air is presented. It is proposed that the green virgin ivy plants naturally available in the environment can be covered along the roofs and walls of the houses of the world. The discussed idea demonstrates the significant potential of the proposed concept of rapidly reducing CO_2 emissions in a relatively shorter time perspective and with relatively lower investment costs, compared to existing other methods. It is estimated that up to 6.89% of the global CO_2 emissions level can be directly reduced by covering the walls and roofs of all houses of the world. Additional advantages of the presented concept are its global nature, as it allows for reduction of CO_2 from all emission sources, regardless of its type and location on earth.

Thus, the proposed idea constitutes one of the applicable technologies allowing to slow down dangerous global warming via dropping global carbon dioxide emissions to zero. Furthermore, it also integrates the efforts of the world population especially from the emerging economies struggling to fulfil their energy needs from fossil fuel-based energy systems with the governments, industrial and scientific communities working on lowering the CO_2 levels in the atmosphere, limiting the global atmospheric temperature rise to 1.5 degrees and fulfilling the global net-zero emissions targets. Therefore, the lowering of the greenhouse gas concentration by covering the walls and roofs of the houses can contribute to the environmental sustainability for future generations, especially in the emerging economies which are struggling with the higher CO_2 emissions levels and the deployment of costly carbon capture technologies could be a challenge for the governments for decarbonizing the industrial sectors.

The technology presented in the paper is particularly valuable for developing countries, as their economies are particularly dependent on fossil fuels, and this technology is simple and inexpensive to implement and allows a rapid reduction in CO_2. However, the presented solutions can be applied in all countries, regardless of their level of development or geographical location.

Future research on the presented technology will be focused on the study of the dynamics of CO_2 adsorption, taking into account various factors affecting the performance of the proposed technology.

Author Contributions: Conceptualisation, J.K.; methodology, J.K. and W.M.A.; software, J.K. and W.M.A.; validation, J.K. and W.M.A.; formal analysis, J.K., W.M.A., T.C., M.S., K.G., A.Z., A.K., D.S., S.M. and Y.G.; investigation, J.K. and W.M.A.; resources, J.K. and W.M.A.; data curation, J.K. and W.M.A.; writing—original draft preparation, J.K. and W.M.A.; writing—review and editing, J.K., W.M.A., T.C., M.S., K.G., A.Z., A.K., D.S., S.M. and Y.G.; visualization, J.K.; supervision, J.K. All authors have read and agreed to the published version of the manuscript.

Funding: The work was performed within the subsidies of the Faculty of Science and Technology, Jan Dlugosz University in Czestochowa, Poland. The support is gratefully acknowledged.

Institutional Review Board Statement: Not applicable.

Informed Consent Statement: Not applicable.

Data Availability Statement: Not applicable.

Conflicts of Interest: The authors declare no conflict of interest.

References

1. Ullah, H.; Fordham, D.A.; Nagelkerken, I. Climate Change Negates Positive CO_2 Effects on Marine Species Biomass and Productivity by Altering the Strength and Direction of Trophic Interactions. *Sci. Total Environ.* **2021**, *801*, 149624. [CrossRef] [PubMed]
2. Zhu, S.; Li, C.; Shao, H.; Ju, W.; Lv, N. The Response of Carbon Stocks of Drylands in Central Asia to Changes of CO_2 and Climate during Past 35 years. *Sci. Total Environ.* **2019**, *687*, 330–340. [CrossRef] [PubMed]
3. Liu, W.-J.; Jiang, H.; Yu, H.-Q. Emerging Applications of Biochar-Based Materials for Energy Storage and Conversion. *Energy Environ. Sci.* **2019**, *12*, 1751–1779. [CrossRef]
4. IEA. Global Energy Review 2021, IEA, Paris. 2021. Available online: https://www.iea.org/reports/global-energy-review-2021 (accessed on 21 January 2022).
5. Friedlingstein, P.; Jones, M.W.; O'Sullivan, M.; Andrew, R.M.; Hauck, J.; Peters, G.P.; Peters, W.; Pongratz, J.; Sitch, S.; Le Quéré, C.; et al. Global Carbon Budget 2019. *Earth Syst. Sci. Data* **2019**, *11*, 1783–1838. [CrossRef]
6. Gates, B. *How to Avoid a Climate Disaster: The Solutions We Have and the Breakthroughs We Need*; Knopf Doubleday Publishing Group: New York, NY, USA, 2021; ISBN 978-0-385-54614-0.
7. Tsai, W.H. Carbon Emission Reduction—Carbon Tax, Carbon Trading, and Carbon Offset. *Energies* **2020**, *13*, 6128. [CrossRef]
8. Board, O.S.; National Academies of Sciences, Engineering, and Medicine. *Negative Emissions Technologies and Reliable Sequestration: A Research Agenda*; National Academies Press: Dulles, VI, USA, 2019.
9. IEA, An Energy Sector Roadmap to Carbon Neutrality in China. 2021. Available online: https://www.iea.org/reports/an-energy-sector-roadmap-to-carbon-neutrality-in-china (accessed on 21 January 2022).
10. International Energy Agency. Net-Zero by 2050 A Roadmap for the Global Energy Sector, Paris. 2021. Available online: https://www.iea.org/reports/net-zero-by-2050 (accessed on 21 January 2022).
11. Ashraf, W.M.; Uddin, G.M.; Arafat, S.M.; Krzywanski, J.; Xiaonan, W. Strategic-level performance enhancement of a 660 MWe supercritical power plant and emissions reduction by AI approach. *Energy Convers. Manag.* **2021**, *250*, 114913. [CrossRef]
12. Abuelgasim, S.; Wang, W.; Abdalazeez, A. A Brief Review for Chemical Looping Combustion as a Promising CO_2 Capture Technology: Fundamentals and Progress. *Sci. Total Environ.* **2021**, *764*, 142892. [CrossRef]
13. Alalwan, H.A.; Alminshid, A.H. CO_2 Capturing Methods: Chemical Looping Combustion (CLC) as a Promising Technique. *Sci. Total Environ.* **2021**, *788*, 147850. [CrossRef]
14. Krzywanski, J.; Żyłka, A.; Czakiert, T.; Kulicki, K.; Jankowska, S.; Nowak, W. A 1.5D Model of a Complex Geometry Laboratory Scale Fuidized Bed Clc Equipment. *Powder Technol.* **2017**, *316*, 592–598. [CrossRef]
15. Muskała, W.; Krzywański, J.; Czakiert, T.; Nowak, W. The Research of CFB Boiler Operation for Oxygen-Enhanced Dried Lignite Combustion. *Rynek Energii* **2011**, *1*, 172–176.
16. Criado, Y.A.; Arias, B.; Abanades, J.C. Calcium Looping CO_2 Capture System for Back-up Power Plants. *Energy Environ. Sci.* **2017**, *10*, 1994–2004. [CrossRef]
17. Krzywanski, J.; Blaszczuk, A.; Czakiert, T.; Rajczyk, R.; Nowak, W. Artificial intelligence treatment of NO_X emissions from CFBC in air and oxy-fuel conditions. CFB-11. In Proceedings of the 11th International Conference on Fluidized Bed Technology, Beijing, China, 14–17 May 2014; pp. 14–17.
18. Zylka, A.; Krzywanski, J.; Czakiert, T.; Idziak, K.; Sosnowski, M.; Grabowska, K.; Prauzner, T.; Nowak, W. The 4th Generation of CeSFaMB in numerical simulations for CuO-based oxygen carrier in CLC system. *Fuel* **2019**, *255*, 115776. [CrossRef]

19. Majchrzak-Kucęba, I.; Wawrzyńczak, D. *Advanced CO_2 Capture Technologies for Clean Coal Energy Generation*; Publishing Office of Czestochowa University of Technology: Częstochowa, Poland, 2016; ISBN 978-83-7193-655-5.
20. Tsai, W.H. Modeling and Simulation of Carbon Emission-Related Issues. *Energies* **2019**, *12*, 2531. [CrossRef]
21. Sukhova, E.M.; Vodeneev, V.A.; Sukhov, V.S. Mathematical Modeling of Photosynthesis and Analysis of Plant Productivity. *Biochem. Moscow Suppl. Ser. A* **2021**, *15*, 52–72. [CrossRef]
22. Lam, M.K.; Lee, K.T.; Mohamed, A.R. Current status and challenges on microalgae-based carbon capture. *Int. J. Greenh. Gas Control.* **2012**, *10*, 456–469. [CrossRef]
23. Mutoti, M.; Gumbo, J.; Jideani, A.I.O. Occurrence of cyanobacteria in water used for food production: A review. *Phys. Chem. Earth* **2022**, *125*, 103101. [CrossRef]
24. Johnson, M.P. Photosynthesis. *Essays Biochem.* **2016**, *60*, 255–273. [CrossRef]
25. Koenig, M. Primitive Dark-Phase Cycle of Photosynthesis at the Origin of Life. *J. Mol. Evol.* **2018**, *86*, 167–171. [CrossRef]
26. Pan, B.T.-C.; Kao, J.-J. Comparison of indices for evaluating building green values based on greenhouse gas emission reductions. *Ecol. Indic.* **2021**, *122*, 107228. [CrossRef]
27. Liu, Q.; Li, F. Spatial and Seasonal Variations of Standardized Photosynthetic Parameters under Different Environmental Conditions for Young Planted *Larix olgensis* Henry Trees. *Forests* **2018**, *9*, 522. [CrossRef]
28. Maiti, R.; Rodriguez, H.G.; Kumari Ch, A. Trees and Shrubs with High Carbon Fixation/Concentration. *For. Res.* **2015**, *S1*, 3. [CrossRef]
29. Kalmatskaya, O.; Karavaev, V.A.; Tikhonov, A.N. Slow induction of chlorophyll a fluorescence excited by blue and red light in *Tradescantia* leaves acclimated to high and low light. *Photosynth. Res.* **2019**, *142*, 265–282. [CrossRef] [PubMed]
30. IBezruk, I.; Marksa, M.; Georgiyants, V.; Ivanauskas, L.; Raudone, L. Phytogeographical profiling of ivy leaf (*Hedera helix* L.). *Ind. Crop. Prod.* **2020**, *154*, 112713. [CrossRef]
31. David, K.; Angier, B. *Foster, Field Guide to Edible Wild Plants*; Messiah College: Mechanicsburg, PA, USA, 2008; ISBN 9780811734479.
32. Steinbrecher, T.; Danninger, E.; Harder, D.; Speck, T.; Kraft, O.; Schwaiger, R. Quantifying the attachment strength of climbing plants: A new approach. *Acta Biomater.* **2010**, *6*, 1497–1504. [CrossRef] [PubMed]
33. United States Department of Agriculture. Plant Guide: Virginia Creeper *Parthenocissus quinquefolia*. Available online: https://plants.usda.gov/DocumentLibrary/factsheet/pdf/fs_paqu2.pdf (accessed on 13 February 2022).
34. Xia, L.; Lenaghan, S.C.; Zhang, M.; Wu, Y.; Zhao, X.; Burris, J.N.; Stewart, C.N. Characterization of English ivy (*Hedera helix*) adhesion force and imaging using atomic force microscopy. *J. Nanopart. Res.* **2011**, *13*, 1029–1037. [CrossRef]
35. MVOrganizing. What Is the Average Size of a Gas Tank? Available online: https://www.mvorganizing.org/what-is-the-average-size-of-a-gas-tank/ (accessed on 16 November 2021).
36. US EPA. Greenhouse Gas Emissions from a Typical Passenger Vehicle. Available online: https://www.epa.gov/greenvehicles/greenhouse-gas-emissions-typical-passenger-vehicle (accessed on 16 November 2021).
37. Fleming, E. How Many Homes Exist in the World?—SidmartinBio. 2021. Available online: https://www.sidmartinbio.org/how-many-homes-exist-in-the-world/#How_many_homes_exist_in_the_world (accessed on 21 January 2022).
38. World Population Clock: 7.9 Billion People (2021)—Worldometer. 2021. Available online: https://www.worldometers.info/world-population/ (accessed on 21 January 2022).
39. Climate Protection with Green Walls—Arguments and Examples. Available online: https://www.fassadengruen.de/en/climate-protection.html (accessed on 26 November 2021).

energies

MDPI

Article

Adsorption of Carbon Dioxide, Methane, and Nitrogen on Zn(dcpa) Metal-Organic Framework

Rui P. P. L. Ribeiro *, Isabel A. A. C. Esteves and José P. B. Mota *

LAQV-REQUIMTE, Department of Chemistry, NOVA School of Science and Technology, NOVA University of Lisbon, 2829-516 Caparica, Portugal; i.esteves@fct.unl.pt

* Correspondence: rpp.ribeiro@fct.unl.pt (R.P.P.L.R.); pmota@fct.unl.pt (J.P.B.M.)

Abstract: Adsorption-based processes using metal-organic frameworks (MOFs) are a promising option for carbon dioxide (CO_2) capture from flue gases and biogas upgrading to biomethane. Here, the adsorption of CO_2, methane (CH_4), and nitrogen (N_2) on Zn(dcpa) MOF (dcpa (2,6-dichlorophenylacetate)) is reported. The characterization of the MOF by powder X-ray diffraction (PXRD), thermogravimetric analysis (TGA), and N_2 physisorption at 77 K shows that it is stable up to 650 K, and confirms previous observations suggesting framework flexibility upon exposure to guest molecules. The adsorption equilibrium isotherms of the pure components (CO_2, CH_4, and N_2), measured at 273–323 K, and up to 35 bar, are Langmuirian, except for that of CO_2 at 273 K, which exhibits a stepwise shape with hysteresis. The latter is accurately interpreted in terms of the osmotic thermodynamic theory, with further refinement by assuming that the free energy difference between the two metastable structures of Zn(dcpa) is a normally distributed variable due to the existence of different crystal sizes and defects in a real sample. The ideal selectivities of the equimolar mixtures of CO_2/N_2 and CO_2/CH_4 at 1 bar and 303 K are 12.8 and 2.9, respectively, which are large enough for Zn(dcpa) to be usable in pressure swing adsorption.

Keywords: Zn(dcpa); MOF; framework flexibility; adsorption; gas storage; biogas; carbon capture

Citation: Ribeiro, R.P.P.L.; Esteves, I.A.A.C.; Mota, J.P.B. Adsorption of Carbon Dioxide, Methane, and Nitrogen on Zn(dcpa) Metal-Organic Framework. *Energies* **2021**, *14*, 5598. https://doi.org/10.3390/en14185598

Academic Editor: Francesco Frusteri

Received: 11 August 2021
Accepted: 3 September 2021
Published: 7 September 2021

1. Introduction

Metal-organic frameworks (MOFs) are being touted as the next generation materials for several adsorptive separation and purification processes [1,2]. MOFs are porous crystalline materials consisting of metal centers connected by organic moieties [3]. An unlimited amount of MOF structures can be envisioned and perhaps synthesized; furthermore, the materials can be tailored for specific applications through pore size tuning and functionalization [4]. These features place MOFs as a very diverse class of materials with potential applications in nearly all fields of chemical engineering [5–12].

Among the available portfolio of MOFs, there are several structures that present structural flexibility, which can be triggered by exposure to specific guest species, changes in temperature or mechanical pressure, or interactions with light or electric fields [13,14]. Framework flexibility generally manifest itself through breathing or gate-opening effects [13]. A comprehensive review regarding MOF flexibility was published by Schneemann et al. [14], in which it is stated that, so far, less than a hundred MOFs have shown important breathing effects. The authors classified the type of flexibility into "breathing", "swelling", "linker rotation", and "subnetwork displacement". The most well-known cases of MOF flexibility are the breathing behavior of MIL-53 [15–17] and the gate-opening of ZIF-8 [18,19].

The MIL-53 family of MOFs consists of trivalent metal (e.g., Al [20], Cr [17], Fe [21], and Sc [22]) terephthalates, which can switch between large-pore (lp) and narrow-pore (np) forms [15–17], with unit cell volume variation of up to 40% [23]. The conformational change can be triggered by different stimuli, for example temperature changes [22,24], application of mechanical pressure [25,26], and adsorption of guest molecules (such as H_2O, CO_2, and other gases) [20,22,26,27]. Interestingly, the synthesis route and solvents employed critically impact the breathing properties of MIL-53 [28], and several authors have observed the absence of the breathing effect in the commercial MIL-53(Al) synthesized by BASF (Basolite©A100) when exposed to CO_2 [29–31].

Another type of MOF flexibility is related to linker rotation [32], of which the most well-known example occurs in ZIF-8, which presents a gate-opening effect [18,19]. The linker rotation triggers a window opening that allows for the adsorption of larger molecules than expected [18].

Recently, MOFs with step-shaped isotherms typical of flexible MOFs have been considered as potential adsorbents for CO_2 capture by temperature swing adsorption (TSA), as they permit decreasing the energy consumption of the process when compared with traditional zeolite 13X systems [33].

Zn(dcpa) is a poorly studied microporous MOF that reportedly exhibits dynamic behavior and stepwise adsorption. Zn(dcpa) is based on paddle-wheel Zn_2 units and unsymmetrical pyridyl dicarboxylate, which give rise to a three-dimensional intersecting pore network with a pore opening of 6.3×12.2 Å^2 [34]. The Zn(dcpa) framework structure is shown in Figure 1. Liu et al. [34] observed the dynamic behavior of Zn(dcpa) upon exposure to N_2 and CO_2 at 77 K and 195 K, respectively. However, they did not observe the MOF's flexible behavior when adsorbing CO_2 at 273 and 293 K and N_2 at 293 K, up to 1 bar.

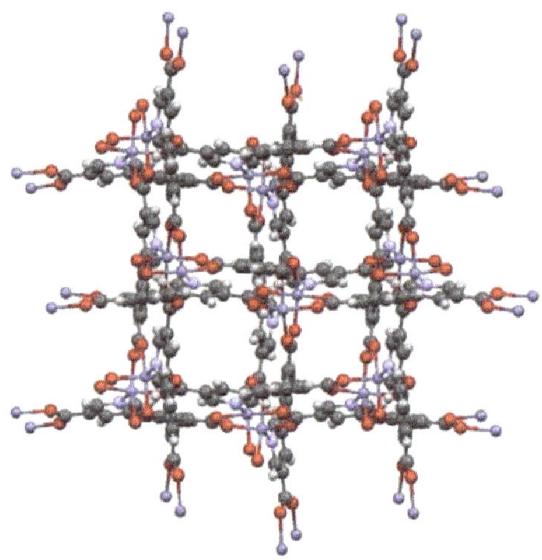

Figure 1. View of the Zn(dcpa) framework along the c-axis (Zn: blue, O: red, C: grey, and H: white). Data generated from the CIF file reported by Liu et al. [34].

In this work, the potential of Zn(dcpa) for application in the separation/purification of gaseous streams containing CO_2, CH_4, and N_2, namely post-combustion CO_2 capture and biogas upgrading, is evaluated. For this purpose, the single-component adsorption equilibria of CO_2, CH_4, and N_2 have been measured at 273–323 K up to 35 bar, and the isosteric heat of adsorption and ideal CO_2/CH_4 and CO_2/N_2 equilibrium selectivities evaluated. Furthermore, the MOF has been characterized regarding its textural properties and thermal stability. The uncommon stepwise adsorption and hysteretic desorption behavior for CO_2 at 273 K has been interpreted in terms of the osmotic thermodynamic theory. The data reported here add important knowledge about the adsorption properties of Zn(dcpa), as prior studies about this MOF are limited to a few publications [34,35].

2. Materials and Methods

2.1. Materials

The Zn(dcpa) MOF sample employed was synthesized at the Materials Center at Technical University Dresden (Germany). After its synthesis, the sample was washed with DMF and, subsequently, activated at 453 K under vacuum for 24 h. The gases employed in the measurements were provided by Air Liquide and Praxair (Portugal) with purities of 99.998% (CO_2), 99.95% (CH_4), 99.99% (N_2), and 99.999% (He).

2.2. Zn(dcpa) Characterization

The sample was characterized using powder X-ray diffraction (PXRD), thermogravimetric analysis (TGA), N_2 physisorption at 77 K, and helium porosimetry. The N_2 adsorption isotherm at 77 K and PXRD were determined by the supplier upon request. TGA analysis was performed using a LABSYS Evo TGA-DTA/DSC from SETARAM Instrumentation, under an argon flow at a heating rate of 3 K/min (up to 1130 K). Helium picnometry was performed at 323 K, in a gravimetric apparatus (described in the next section), to determine the skeletal density of the MOF (ρ_s).

2.3. Single-Component Adsorption Equilibrium

Single-component adsorption equilibrium isotherms of CO_2, CH_4, and N_2 at 273 K, 303 K, and 323 K, between 0 and 35 bar, were determined using the standard static gravimetric method [31,36,37]. The measurements were performed in a high-pressure magnetic-suspension balance ISOSORP 2000 (Rubotherm GmbH, Germany) using approximately 600 mg of Zn(dcpa) powder. Both the adsorption and desorption data were recorded to evaluate the hysteretic effects. The sample was received from the supplier already activated and stored in an argon atmosphere, which is why the pre-treatment performed before measuring the adsorption equilibrium isotherms was limited to overnight vacuum. The experimental setup and procedure are detailed elsewhere [31,37].

The excess amount adsorbed, q_{exc}, is determined as follows

$$q_{exc} = \frac{w - m_s - m_h + V_h \rho_g}{m_s} + v_s \rho_g \quad (1)$$

where w is the apparent mass weighted; m_s is the mass of MOF; V_h and m_h correspond to the volume and mass of the measuring cell, respectively, which contribute to the buoyancy effects; ρ_g is the density of the bulk gas at the experimental conditions; and v_s is the specific volume of the solid matrix of the MOF ($v_s = 1/\rho_s$, where ρ_s is the skeletal density of the adsorbent). v_s was determined by helium pycnometry performed at 323 K in the gravimetric apparatus. This was determined assuming that He penetrates the MOF pore volume without being adsorbed.

The absolute amount adsorbed, q, can be determined from the excess amount adsorbed, using the following

$$q = q_{exc}\left(\frac{\rho_l}{\rho_l - \rho_g}\right) \quad (2)$$

assuming that the adsorbed phase density corresponds to the density of the liquid at its boiling point at 1 atm (ρ_l) [38].

3. Results and Discussion

3.1. Zn(dcpa) Characterization

The PXRD patterns obtained are displayed in Figure 2, showing that the position of the reflexes changed during activation. This effect was also observed by Liu et al. [34], who associated it with framework shrinkage upon removal of the guest molecules. The authors also observed the reversibility of this phenomenon when readsorbing the solvent.

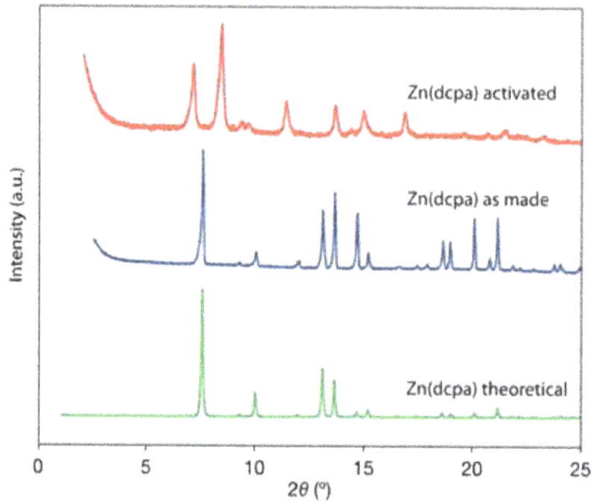

Figure 2. PXRD diffraction pattern for activated, as made, and theoretical Zn(dcpa) samples.

The thermal stability of the sample was also characterized by TGA; the recorded sample mass as a function of the heating temperature is shown in Figure 3. The results show an initial mass decrease (~10%) from room temperature up to around 400 K, due to the removal of pre-adsorbed impurities and humidity, due to MOF exposure to the indoor atmosphere just prior to the analysis. MOF was stable up to 650 K, after which a steep decrease in the mass was observed, reaching a mass decrease of 50%, similar to the behavior observed by Liu et al. [34]. Above 750 K, the Zn(dcpa) mass decreased more smoothly until reaching the remaining experimental amount of ca. 25% at 1130 K.

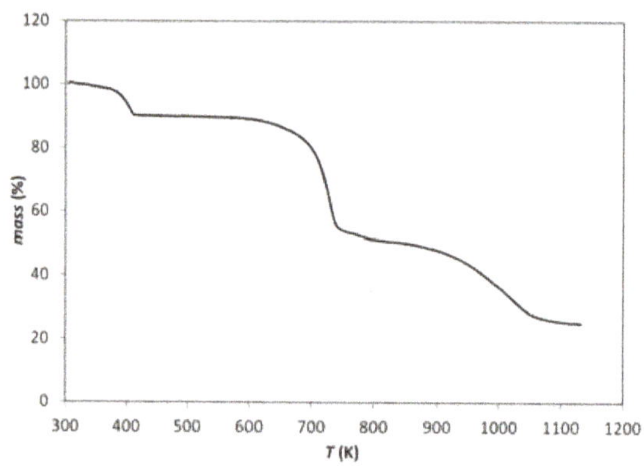

Figure 3. TGA results for Zn(dcpa) as a function of temperature (heating rate of 3 K/min).

The Zn(dcpa) porosity was evaluated by N_2 adsorption at 77 K. The obtained isotherm is plotted in Figure 4, showing an initial step followed by a smoother increase until $p/p_0 = 0.06$; then, another steep increase is observed before reaching a nearly constant plateau with only a small increase between $p/p_0 = 0.2$ (274 cm^3/g) and $p/p_0 = 0.97$ (303 cm^3/g). Note that p and p_0 are the equilibrium and saturation pressures of the adsorbate at 77 K, respectively. The same behavior was observed by Liu et al. [34] who attributed the first step of the isotherm to the Zn(dcpa) structure with shrunken pores, and the second step to an expanded structure. In our work, the expanded structure had a specific pore volume of 0.47 cm^3/g, determined at a relative pressure of $p/p_0 = 0.97$, assuming the pores were filled with condensed liquid N_2 at its normal boiling point. The desorption branch showed hysteresis at $p/p_0 < 0.2$, which is also in accordance with a previous report, although in our case, the hysteresis loop seemed to close at lower pressures—which corresponds to a return to the shrunken pore conformation—as opposed to the observation of Liu et al. [34].

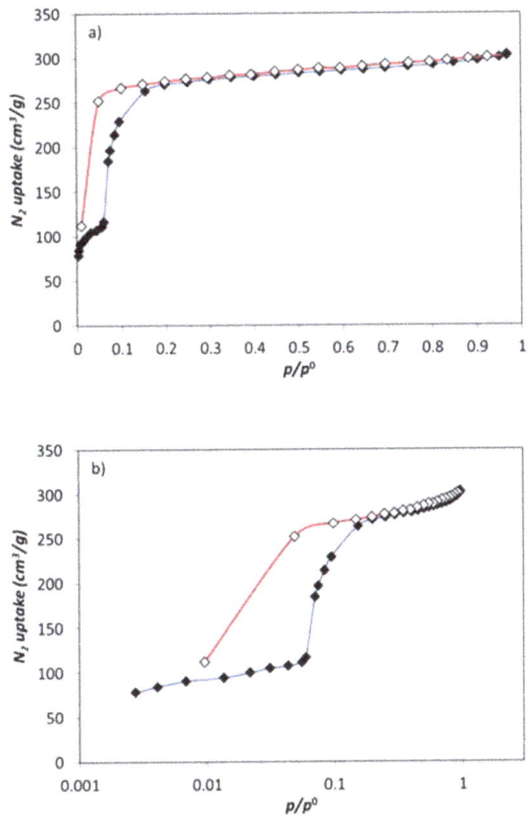

Figure 4. N_2 adsorption equilibrium isotherm at 77 K on Zn(dcpa) in linear (**a**) and log (**b**) scales. Filled and empty symbols represent the adsorption and desorption data, respectively.

3.2. *Single-Component Adsorption Equilibrium*

Prior to the adsorption of CO_2, CH_4, and N_2, the skeletal density of Zn(dcpa) was determined by helium picnometry at 323 K, obtaining a $\rho_s = 1.74\ \text{g/cm}^3$ ($v_s = 1/\rho_s = 0.575\ \text{cm}^3/\text{g}$). For a purely crystalline porous material with a regular lattice, $v_\mu + v_s$ is equal to the specific volume of the unit cell of the lattice. The particle density determined, $\rho_p = 1/(v_s + v_\mu) = 0.957\ \text{g/cm}^3$, is in excellent agreement with the value obtained from the crystallographic data ($\rho_p = 0.961\ \text{g/cm}^3$) by Liu et al. [34].

The adsorption equilibria of CO_2, CH_4, and N_2 on the Zn(dcpa) MOF were measured at 273, 303, and 323 K over the pressure range of 0 to 35 bar. The CO_2, CH_4, and N_2 absolute adsorption equilibrium isotherms obtained are reported in Figures 5–7, respectively. The CO_2 adsorption isotherms were quite steep in the Henry region, showing a high adsorption capacity at a low pressure, an important feature for use in post-combustion CO_2 capture applications. On the other hand, the N_2 adsorption isotherms were much more linear and had lower adsorption capacity; the CH_4 adsorption isotherms were intermediate between those of CO_2 and N_2.

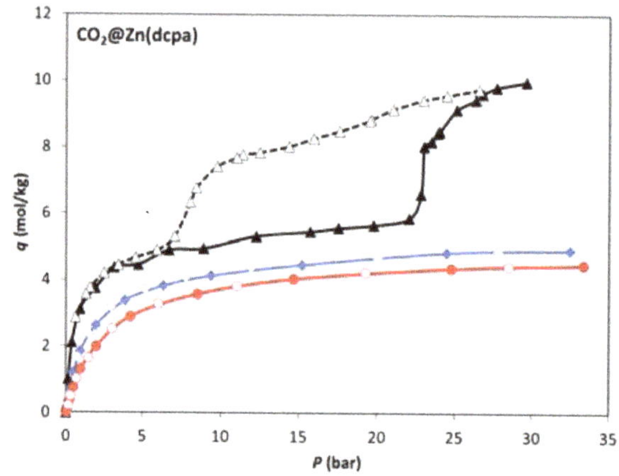

Figure 5. Absolute adsorption equilibrium isotherms of CO_2 on Zn(dcpa) at 273 K (▲), 303 K (♦), and 323 K (•). Filled and empty symbols represent adsorption and desorption experimental data, respectively.

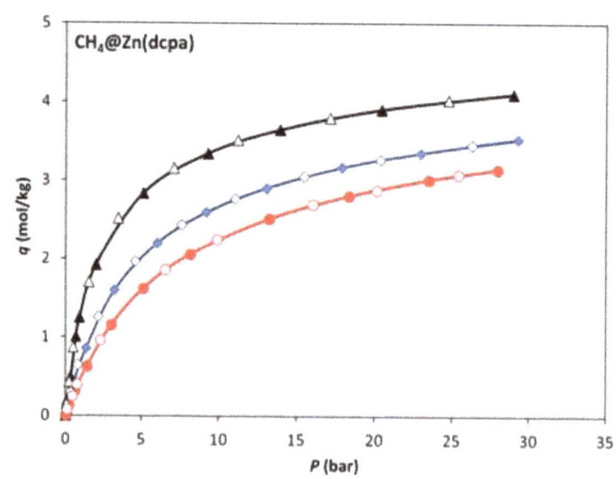

Figure 6. Absolute adsorption equilibrium isotherms of CH_4 on Zn(dcpa) at 273 K (▲), 303 K (♦), and 323 K (•). Filled and empty symbols represent adsorption and desorption experimental data, respectively.

An interesting feature of the CO_2 adsorption equilibrium isotherms can be observed in Figure 5. At 273 K, the adsorption branch of the isotherm follows a typical Langmuir-type shape up to approximately 22 bar, where a step in the isotherm is observed. This behaviour is similar to that observed by Liu et al. [34] for CO_2 adsorption at 195 K, which the authors related to the transition between a shrunken-pore phase and an expanded-pore one. The desorption branch then follows a different path than the adsorption one, showing a hysteresis loop that closes at 7 bar. The reproducibility of this behaviour was checked by repeating the measurements. The stepwise CO_2 adsorption observed is an interesting feature of Zn(dcpa) that can be explored for gas separation or storage applications. It can enhance the working capacity of the solid material upon mild pressure or temperature swings [33]. Despite the observation of the MOF flexibility for CO_2 adsorption at 273 K, the same behaviour was not observed for any of the other temperatures nor for the adsorbate species tested (CH_4 and N_2).

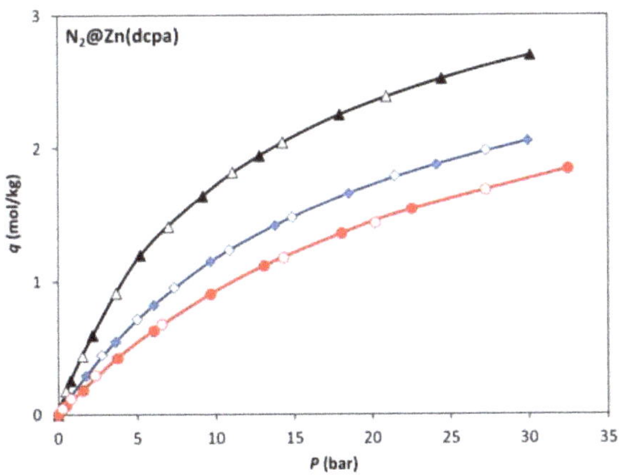

Figure 7. Absolute adsorption equilibrium isotherms of N_2 on Zn(dcpa) at 273 K (▲), 303 K (♦), and 323 K (•). Filled and empty symbols represent adsorption and desorption experimental data, respectively.

3.3. Osmotic Thermodynamic Theory

Some materials present clear transitions between different metastable framework structures. Zn(dcpa) is an example of such materials. In these cases, an "osmotic subensemble" [15,39–42] can be employed to describe the equilibrium between host structures when exposed to gaseous adsorbates. Alternatively, Ghysels et al. [42] proposed another free energy model able to describe the thermodynamics of breathing phenomena in flexible materials. In this work, we interpreted our data using the former approach.

The osmotic potential [43] of the solid–adsorbate system for the ith structure of Zn(dcpa), either shrunken-pore (SP) or expanded-pore (EP), is

$$\begin{aligned}\Omega_{\mathrm{os}}^{(i)}(P,T) &= F_{\mathrm{host}}^{(i)}(T) + Pv_{\mathrm{p}}^{(i)} - \int_0^P q^{(i)}(P,T)v_{\mathrm{g}}(P,T)\,dP \\ &= F_{\mathrm{host}}^{(i)}(T) + Pv_{\mathrm{p}}^{(i)} - RT\int_0^P q^{(i)}(P,T)Z(P,T)\,d\ln P\end{aligned} \tag{3}$$

where $F_{\mathrm{host}}^{(i)}(T)$ corresponds to the empty structure's free energy at temperature T and $v_{\mathrm{p}}^{(i)}$ to its apparent specific volume (i.e., to the sum of its skeletal, v_{s}, and porous, $v_{\mu}^{(i)}$, volumes); $q^{(i)}(P,T)$ corresponds to the adsorption isotherm considering a rigid framework in its ith structural form; and $v_{\mathrm{g}} = 1/n_{\mathrm{g}}$ and P are the molar volume and compressibility factor of the adsorptive. The difference in the osmotic potential between both of the structures considered (EP and SP), $\Delta\Omega_{\mathrm{os}}(P,T) = \Omega_{\mathrm{os}}^{(\mathrm{EP})}(P,T) - \Omega_{\mathrm{os}}^{(\mathrm{SP})}(P,T)$, is thus

$$\Delta\Omega_{\mathrm{os}}(P,T) = \Delta F_{\mathrm{host}}(T) + P\Delta v_{\mathrm{p}} - RT\int_0^P \Delta q(P,T)Z(P,T)\,d\ln P \tag{4}$$

where $\Delta\phi \equiv \phi^{(\mathrm{EP})} - \phi^{(\mathrm{SP})}$ is the difference in the value of property ϕ between the EP and SP structures at temperature T. If $\Delta\Omega_{\mathrm{os}} > 0$, the SP structure will be more stable than EP; if $\Delta\Omega_{\mathrm{os}} < 0$, the reverse will be true. For Zn(dcpa), $v_{\mu}^{(\mathrm{EP})} = 0.47\ \mathrm{cm^3/g}$ and $v_{\mu}^{(\mathrm{SP})} = 0.14\ \mathrm{cm^3/g}$; hence $\Delta v_{\mathrm{p}} = (v_{\mathrm{s}} + v_{\mu})^{(\mathrm{EP})} - (v_{\mathrm{s}} + v_{\mu})^{(\mathrm{SP})} \approx v_{\mu}^{(\mathrm{EP})} - v_{\mu}^{(\mathrm{SP})} = 0.33\ \mathrm{cm^3/g}$. Assuming ideal gas behavior ($Z \approx 1$), the previous equation can be simplified to

$$\Delta\Omega_{\mathrm{os}}(P,T) \approx \Delta F_{\mathrm{host}}(T) + P\Delta v_{\mathrm{p}} - RT\int_0^P \Delta q(P,T)\,d\ln P. \tag{5}$$

If ΔF_{host} and $\Delta q(P)$ are known at a given temperature T, putting $\Delta\Omega_{\mathrm{os}} = 0$ in Equation (4) (or Equation (5)) and solving it for P gives the pressure at which the phase transition occurs at T. However, in real scenarios, the MOF crystals have defects and the sample has a distribution of crystal sizes, both contributing to smoothing the structural transitions between the two metastable framework structures. Here, we extend the osmotic thermodynamic theory to account for this diffuse effect. It is assumed that for a real sample, ΔF_{host} is normally (Gaussian) distributed around the corresponding value for a perfect crystal with a probability density function

$$f(\Delta F_{\mathrm{host}}) = \frac{1}{\sigma_{\Delta F}\sqrt{2\pi}} \exp\left(-\frac{1}{2}\frac{(\Delta F_{\mathrm{host}} - \Delta F_{\mu})^2}{\sigma_{\Delta F}^2}\right), \tag{6}$$

where ΔF_{μ} is the mean or expectation of the distribution (i.e., the value of ΔF_{host} for a perfect crystal) and $\sigma_{\Delta F}$ is its standard deviation. Given that in the case under study, the structural transition is triggered by exposure to a specific guest species, the previous hypothesis is almost equivalent to considering that the adsorptive pressure, P, that triggers the structural transition at a fixed temperature is also a normally distribute variable and, therefore, that its cumulative distribution function, $\Phi(P)$, at a fixed temperature is

$$\Phi(P) \equiv \int f(P)dP = \frac{1}{2}\left[1 + \mathrm{erf}\left(\frac{P - P_{\mu}}{\sigma_P\sqrt{2}}\right)\right] \tag{7}$$

where P_{μ} is the mean or expectation of the distribution (i.e., the transition pressure for a perfect crystal) and σ_P is its standard deviation. Note that $\Phi(x) = \mathrm{prob}(P \leq x)$, where the right-hand side represents the probability that P takes on a value less than or equal to x. Therefore, the macroscopically observed adsorption branch of the isotherm for a real sample is given by

$$\begin{matrix} q_{\mathrm{ads}}(P) = [1 - \Phi_{\mathrm{ads}}(P)]q^{(\mathrm{SP})}(P) + \Phi_{\mathrm{ads}}(P)q^{(\mathrm{EP})}(P) \\ q_{\mathrm{des}}(P) = [1 - \Phi_{\mathrm{des}}(P)]q^{(\mathrm{SP})}(P) + \Phi_{\mathrm{des}}(P)q^{(\mathrm{EP})}(P) \end{matrix} \quad (\text{fixed } T) \tag{8}$$

where $q^{(i)}(P,T)$ is the adsorption isotherm considering a rigid framework constrained to its ith form.

This model was fitted to our experimental data, assuming the adsorption isotherms for the metastable forms of the framework are Langmuirian, i.e.,

$$q^{(i)}(P) = \frac{q_{\infty}^{(i)} b^{(i)} P}{1 + b^{(i)} P} \tag{9}$$

where $q_{\infty}^{(i)}$ and $b^{(i)}$ are the saturation capacity and equilibrium constant for the ith form, respectively, in which case Equation (5) reduces to

$$\Delta\Omega_{\mathrm{os}}(P,T) = \Delta F_{\mathrm{host}}(T) + P\Delta v_{\mathrm{p}} - RT\left[q_{\infty}^{(\mathrm{EP})}\ln\left(1 + b^{(\mathrm{EP})}P\right) - q_{\infty}^{(\mathrm{SP})}\ln\left(1 + b^{(\mathrm{SP})}P\right)\right] \tag{10}$$

Table 1 lists the parameter values resulting from the model fitting to the experimental data, saturation capacity and Langmuir equilibrium constant ($q_{\infty}^{(i)}$ and $b^{(i)}$) for CO_2 adsorption at 273 K in the SP and EP metastable structures of Zn(dcpa), mean and standard deviation (P_{μ} and σ_{μ},) of the Gaussian distribution of adsorptive pressure that triggers the phase transition along the adsorption and desorption branches of the isotherm, and the corresponding free energy changes of the empty structure (ΔF_{host}).

Table 1. Fitted parameters of the osmotic thermodynamic theory to the adsorption (Ads) and desorption (Des) branches of the experimental CO2 isotherm at 273 K. $\Delta F_{\text{host}} \equiv F^{(\text{EP})}_{\text{host}} - F^{(\text{SP})}_{\text{host}}$.

	q_∞ **(mol/kg)**	b **(atm $^{-1}$)**		P_μ **(atm)**	σ_P **(atm)**	ΔF_{host} **(J/g)**
SP	5.65	1.30	Ads	23.0	1.3	−4.45
EP	12.2	0.143	Des	7.0	0.5	−10.7

Figure 8 compares the experimental adsorption data and the fitted osmotic thermodynamic model, and shows that excellent agreement with the experimental results has been reached using the proposed procedure, substantiated by the fact that the plotted adsorption and desorption curves accurately reproduce the experimental data.

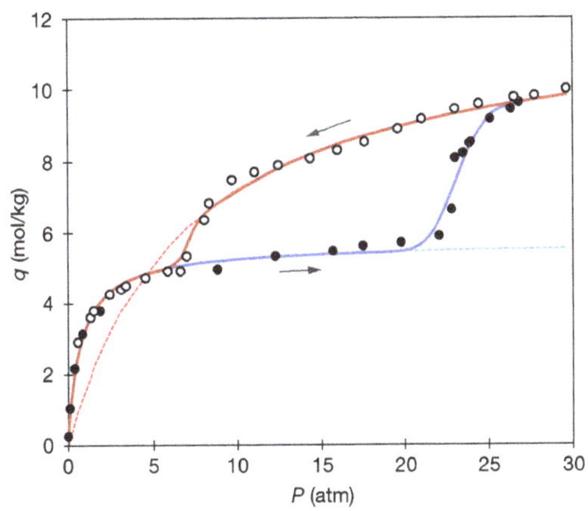

Figure 8. Fitting of the osmotic thermodynamic theory to the adsorption (•) and desorption (○) branches of the experimental CO_2 isotherm at 273 K. Dashed lines: fitted Langmuir adsorption isotherms if the framework were rigid and constrained to its EP (- - -) and SP (- - -) forms; solid lines: predicted adsorption (——) and desorption (——) branches of the isotherm at 273 K.

The first two numeric columns of Table 2 list the parameter values of the Langmuir adsorption isotherm model that best fit the CH_4 and N_2 experimental adsorption data at 273 K (Figure 9); these values apply when the framework is in its SP form. The last two columns of Table 2 list the corresponding values if the framework were hypothetically in EP form; given the absence of experimental data, the Langmuir parameters were estimated using the following scaling rules:

$$q^{(\text{EP})}_{i,\infty} = q^{(\text{SP})}_{i,\infty} \frac{q^{(\text{EP})}_{CO_2,\infty}}{q^{(\text{SP})}_{CO_2,\infty}} \text{ and } b^{(\text{EP})}_i = b^{(\text{SP})}_i \frac{b^{(\text{EP})}_{CO_2}}{b^{(\text{SP})}_{CO_2}} \tag{11}$$

Although these rules are rather crude and their application cannot be considered quantitatively precise, they allow us to explain why the adsorption isotherms of CH_4 and N_2 do not point to a phase transition between the SP and EP structures (and no hysteresis) in the pressure range tested experimentally by us. The reason is that in the case of CH_4 or N_2 adsorption, the pressure must be increased considerably, well above the maximum experimental value tested by us, for the term $RT \int_0^P \Delta q(P,T)\, d\ln P$ to change the sign of $\Delta\Omega_{\text{os}}$. A similar reasoning explains why the CO_2 adsorption isotherms at 303 K and 323 K do not hint at a phase transition between the SP and EP structures.

Table 2. Fitting of the Langmuir adsorption isotherm model to the CH_4 and N_2 experimental adsorption data at 273 K. The parameters for the EP form are estimated as $q_{i,\infty}^{(EP)} = q_{i,\infty}^{(SP)} q_{CO_2,\infty}^{(EP)} / q_{CO_2,\infty}^{(SP)}$ and $b_i^{(EP)} = b_i^{(SP)} b_{CO_2}^{(EP)} / b_{CO_2}^{(SP)}$.

	$q_\infty^{(SP)}$ (mol/kg)	$b^{(SP)}$ (atm $^{-1}$)	$q_\infty^{(EP)}$ (mol/kg)	$b^{(EP)}$ (atm $^{-1}$)
N_2	3.681	0.089	7.948	0.010
CH_4	4.328	0.411	9.346	0.045

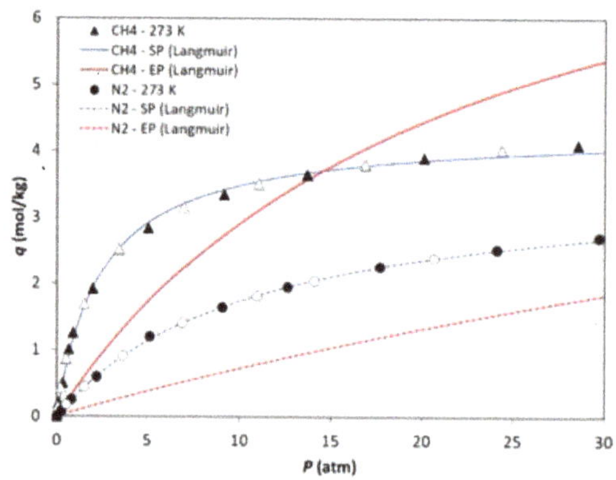

Figure 9. Fitting of the Langmuir isotherm model to the CH_4 (▲) and N_2 (•) experimental adsorption equilibrium isotherms at 273 K. Filled and empty symbols denote experimental adsorption and desorption data, respectively. Solid lines: fitted CH_4 Langmuir isotherm for rigid framework constrained to its EP (red) and SP (blue) forms; dashed lines: fitted N_2 Langmuir isotherm for rigid framework constrained to its EP (red) and SP (blue) forms.

3.4. *Potential of Zn(dcpa) for CO_2/N_2 and CO_2/CH_4 Separation*

To assess the potential use of Zn(dcpa) for the adsorptive separation of CO_2/N_2 and CO_2/CH_4, we first compared the single-component equilibrium isotherms and the obtained selectivities. Figure 10a compares the isotherms obtained at 303 K for the three gases. The ideal selectivity for an equimolar mixture, $\alpha_{A/B} = q_A/q_B$, was calculated using the single-component adsorption capacity ratio; the obtained results are reported in Figure 10b,c for the CO_2/N_2 and CO_2/CH_4 selectivities, respectively. Both the CO_2/N_2 and CO_2/CH_4 selectivities decreased with the increasing pressure, ranging from 12.8 (at 1 bar) to 6.7 (6 bar) for CO_2/N_2, and from 2.9 (at 1 bar) to 2.1 (6 bar) for CO_2/CH_4. The CO_2/N_2 selectivity was significantly higher than that of CO_2/CH_4, especially at a lower pressure. Although the reported equilibrium selectivity did not take into account the influence of adsorption kinetics or the impact of a real gas mixture, these results serve as a first evaluation of the good potential of Zn(dcpa) for CO_2 separation from CO_2/N_2 and CO_2/CH_4 mixtures. Comparing the selectivity of Zn(dcpa) for CO_2/N_2 with those of the commercial MOFs MIL-53(Al) [31], ZIF-8, [44], and Fe-BTC [45] (Figure 10b,c), it is concluded that the former outperformed the others at a lower pressure. For example, at 1 bar, the order of selectivities was 12.8 (Zn(dcpa)) > 10.5 (Fe-BTC) > 9.6 (MIL-53(Al)) > 5.9 (ZIF-8). The same can be said about the CO_2/CH_4 selectivity trend: 2.9 (Zn(dcpa)) > 2.7 (MIL-53(Al)) > 2.6 (ZIF-8), although in this case, ZIF-8 surpassed Zn(dcpa) at pressures above 2.4 bar.

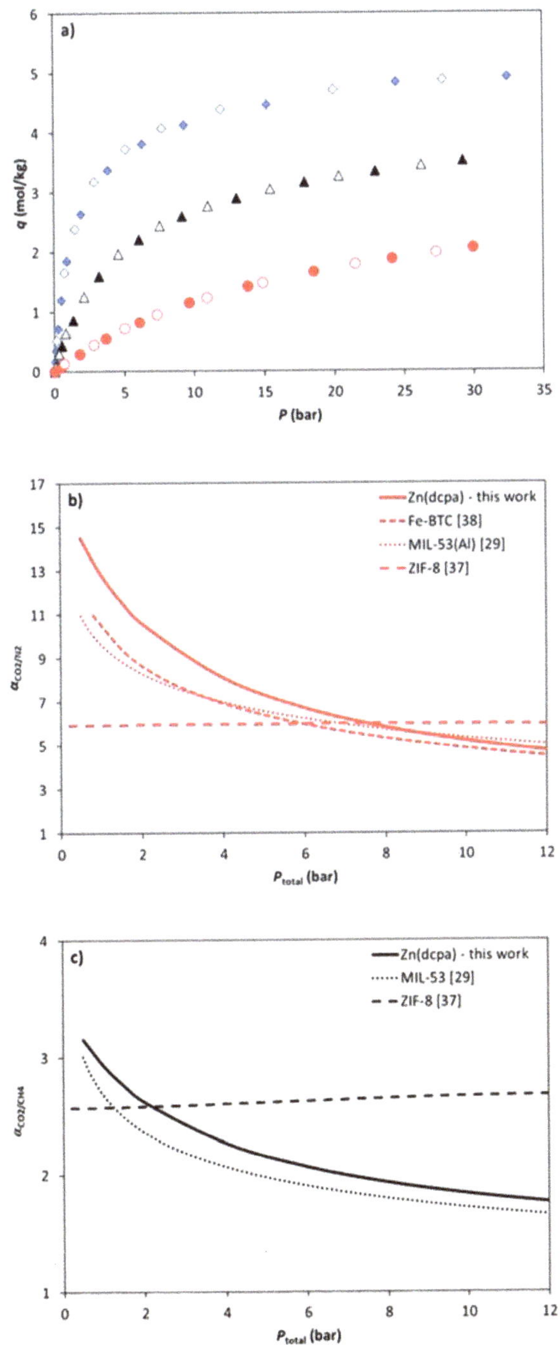

Figure 10. (**a**) Adsorption equilibrium isotherms of CO_2 (♦), CH_4 (▲), and N_2 (•) on Zn(dcpa) at 303 K. Filled and empty symbols represent adsorption and desorption data, respectively; (**b**) CO_2/N_2 and (**c**) CO_2/CH_4 equilibrium selectivity at 303 K as a function of pressure for Zn(dcpa) and commercial MOFs MIL-53(Al) [31], ZIF-8 [44], and Fe-BTC [45].

The isosteric heat of adsorption, Q_{st}, was determined from the experimental data via the Clausius–Clapeyron equation: $(\log P)_q = \text{const} - Q_{st}/RT$ [46] Using this approach, the plot of log P versus $-1/RT$, at constant loading, should give a straight line from which the slope of Q_{st} can be obtained. The plots in Figure 11 of Q_{st} for CO_2, CH_4, and N_2 on Zn(dcpa) as a function of loading show that for CH_4, their values were approximately constant within the loading range studied, while for CO_2 and N_2, a linear increase was observed. It should be noted that in the case of CO_2, Q_{st} was plotted for loadings lower than the MOF's conformational change observed at 273 K, i.e., less than 4 mol/kg. The isosteric heat of the adsorption was higher for CO_2 (23–28 kJ/mol), which is in accordance with the values reported in the literature [34], followed by that for CH_4 (~19.5 kJ/mol) and N_2 (13–15 kJ/mol).

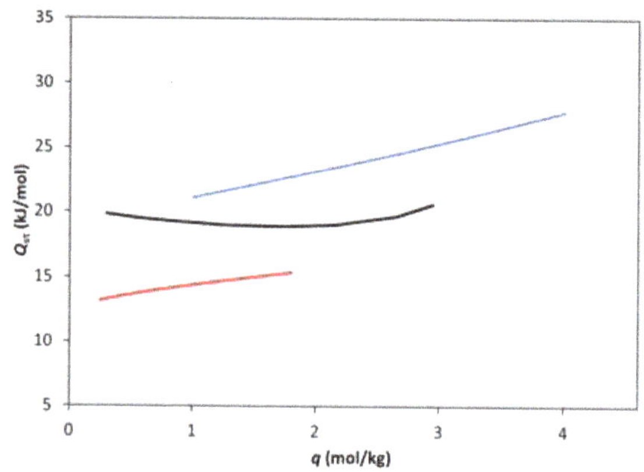

Figure 11. CO_2 (blue), CH_4 (black), and N_2 (red) isosteric heats of adsorption, Q_{st}, as a function of loading, q.

4. Conclusions

Zn(dcpa) MOF was characterized through PXRD, thermogravimetric analysis, and N_2 adsorption at 77 K. In line with the XRD and N_2 data reported by Liu et al. [34], our results indicate that the framework conformation changes (pore shrinkage/pore expansion) upon removal/loading of guest molecules. The TGA results demonstrate that the MOF is stable up to 650 K.

The adsorption equilibrium isotherms of CO_2, CH_4, and N_2 on Zn(dcpa) at 273 K, 303 K, and 323 K are reported up to 35 bar. The obtained data highlight the interesting behavior of CO_2 adsorption at 273 K, which exhibits a stepped isotherm related to the transition between shrunken- and expanded-pore phases at around 22 bar. When observing the desorption branch of the isotherm, a hysteresis loop is present, closing at 7 bar. Although the same behavior is observed for CO_2 adsorption at 195 K [34], this is the first report of this effect at 273 K. None of the remaining CO_2 (303 K and 323 K), CH_4, or N_2 isotherms show the same behavior.

The CO_2 adsorption equilibrium at 273 K is accurately interpreted using the osmotic thermodynamic theory, which is further refined by considering that the free energy difference between the two metastable structures of Zn(dcpa) is a normally distributed variable due to the distribution of crystal sizes and the defects in a real MOF sample.

Regarding the uptake amounts, Zn(dcpa) can adsorb higher amounts of CO_2, followed by CH_4 and N_2. The CO_2/N_2 and CO_2/CH_4 ideal equilibrium selectivities of Zn(dcpa) at 303 K are evaluated for equimolar mixtures, resulting in 12.8 and 2.9, respectively, for a total pressure of 1 bar.

The reported data are essential for the modelling of adsorption-based processes, namely pressure swing adsorption (PSA) and temperature swing adsorption (TSA), for the separation of mixtures containing the studied gases, e.g., biogas upgrading and CO_2 capture from flue gases.

Author Contributions: Conceptualization, R.P.P.L.R. and J.P.B.M.; formal analysis, R.P.P.L.R. and J.P.B.M.; investigation, R.P.P.L.R. and I.A.A.C.E.; writing—original draft preparation, R.P.P.L.R.; writing—review and editing, R.P.P.L.R., I.A.A.C.E. and J.P.B.M.; visualization, R.P.P.L.R. and J.P.B.M. All authors have read and agreed to the published version of the manuscript.

Funding: This research was funded by the Associate Laboratory for Green Chemistry (AQV), which is financed by national funds from FCT/MCTES (UIDB/50006/2020 and UIDP/50006/2020). Rui Ribeiro and Isabel Esteves acknowledge financial support from FCT/MCTES through the Norma Transitória DL 57/2016 Program Contract and project IF/01016/2014, respectively. The authors also acknowledge support from the ERANet LAC initiative through project ELAC2014/BEE0367.

Conflicts of Interest: The authors declare no conflict of interest. The funders had no role in the design of the study; in the collection, analyses, or interpretation of data; in the writing of the manuscript; or in the decision to publish the results.

References

1. Herm, Z.R.; Swisher, J.A.; Smit, B.; Krishna, R.; Long, J.R. Metal-organic frameworks as adsorbents for hydrogen purification and precombustion carbon dioxide capture. *J. Am. Chem. Soc.* **2011**, *133*, 5664–5667. [CrossRef] [PubMed]
2. Grande, C.A.; Blom, R.; Andreassen, K.A.; Stensrød, R.E. Experimental results of pressure swing adsorption (PSA) for pre-combustion CO_2 capture with metal organic frameworks. *Energy Procedia* **2017**, *114*, 2265–2270. [CrossRef]
3. Rowsell, J.L.C.; Yaghi, O.M. Metal–organic frameworks: A new class of porous materials. *Micropor. Mesopor. Mat.* **2004**, *73*, 3–14. [CrossRef]
4. Ferey, G. Hybrid porous solids: Past, present, future. *Chem. Soc. Rev.* **2008**, *37*, 191–214. [CrossRef] [PubMed]
5. Millward, A.R.; Yaghi, O.M. Metal-organic frameworks with exceptionally high capacity for storage of carbon dioxide at room temperature. *J. Am. Chem. Soc.* **2005**, *127*, 17998–17999. [CrossRef]
6. Furukawa, H.; Cordova, K.E.; O'Keeffe, M.; Yaghi, O.M. The chemistry and applications of metal-organic frameworks. *Science* **2013**, *341*. [CrossRef]
7. Czaja, A.U.; Trukhan, N.; Muller, U. Industrial applications of metal-organic frameworks. *Chem. Soc. Rev.* **2009**, *38*, 1284–1293. [CrossRef] [PubMed]
8. Meek, S.T.; Greathouse, J.A.; Allendorf, M.D. Metal-organic frameworks: A rapidly growing class of versatile nanoporous materials. *Adv. Mater.* **2011**, *23*, 249–267. [CrossRef]
9. Mueller, U.; Schubert, M.; Teich, F.; Puetter, H.; Schierle-Arndt, K.; Pastre, J. Metal-organic frameworks—Pospective industrial applications. *J. Mater. Chem.* **2006**, *16*, 626–636. [CrossRef]
10. Ribeiro, R.P.P.L.; Antunes, C.L.; Garate, A.U.; Portela, A.F.; Plaza, M.G.; Mota, J.P.B.; Esteves, I.A.A.C. Binderless shaped metal-organic framework particles: Impact on carbon dioxide adsorption. *Microporous Mesoporous Mater.* **2019**, *275*, 111–121. [CrossRef]
11. An, Y.; Tian, Y.; Li, Y.; Wei, C.; Tao, Y.; Liu, Y.; Xi, B.; Xiong, S.; Feng, J.; Qian, Y. Heteroatom-doped 3D porous carbon architectures for highly stable aqueous zinc metal batteries and non-aqueous lithium metal batteries. *Chem. Eng. J.* **2020**, *400*, 125843. [CrossRef]
12. An, Y.; Tian, Y.; Li, Y.; Xiong, S.; Zhao, G.; Feng, J.; Qian, Y. Green and tunable fabrication of graphene-like N-doped carbon on a 3D metal substrate as a binder-free anode for high-performance potassium-ion batteries. *J. Mater. Chem. A* **2019**, *7*, 21966–21975. [CrossRef]
13. Alhamami, M.; Doan, H.; Cheng, C.-H. A review on breathing behaviors of metal-organic-frameworks (MOFs) for gas adsorption. *Materials* **2014**, *7*, 3198–3250. [CrossRef]
14. Schneemann, A.; Bon, V.; Schwedler, I.; Senkovska, I.; Kaskel, S.; Fischer, R.A. Flexible metal-organic frameworks. *Chem. Soc. Rev.* **2014**, *43*, 6062–6096. [CrossRef]
15. Coudert, F.-X.; Mellot-Draznieks, C.; Fuchs, A.H.; Boutin, A. Double structural transition in hybrid material MIL-53 upon hydrocarbon adsorption: The termodynamics behind the scenes. *J. Am. Chem. Soc.* **2009**, *131*, 3442–3443. [CrossRef] [PubMed]
16. Mishra, P.; Edubilli, S.; Uppara, H.P.; Mandal, B.; Gumma, S. Effect of adsorbent history on adsorption characteristics of MIL-53 (Al) metal organic framework. *Langmuir* **2013**, *29*, 12162–12167. [CrossRef] [PubMed]
17. Serre, C.; Millange, F.; Thouvenot, C.; Noguès, M.; Marsolier, G.; Louër, D.; Férey, G. Very large breathing effect in the first nanoporous chromium (III)-based solids: MIL-53 or Cr^{III} (OH)·$\{O_2C-C_6H_4-CO_2\}$·$\{HO_2C-C_6H_4-CO_2H\}_x$·$H2O_y$. *J. Am. Chem. Soc.* **2002**, *124*, 13519–13526. [CrossRef]
18. Fairen-Jimenez, D.; Galvelis, R.; Torrisi, A.; Gellan, A.D.; Wharmby, M.T.; Wright, P.A.; Mellot-Draznieks, C.; Duren, T. Flexibility and swing effect on the adsorption of energy-related gases on ZIF-8: Combined experimental and simulation study. *Dalton Trans.* **2012**, *41*, 10752–10762. [CrossRef] [PubMed]

19. Fairen-Jimenez, D.; Moggach, S.A.; Wharmby, M.T.; Wright, P.A.; Parsons, S.; Düren, T. Opening the gate: Framework flexibility in ZIF-8 explored by experiments and simulations. *J. Am. Chem. Soc.* **2011**, *133*, 8900–8902. [CrossRef]
20. Loiseau, T.; Serre, C.; Huguenard, C.; Fink, G.; Taulelle, F.; Henry, M.; Bataille, T.; Ferey, G. A rationale for the large breathing of the porous aluminum terephthalate (MIL-53) upon hydration. *Chem. Eur. J.* **2004**, *10*, 1373–1382. [CrossRef] [PubMed]
21. Hamon, L.; Serre, C.; Devic, T.; Loiseau, T.; Millange, F.; Férey, G.; Weireld, G.D. Comparative study of hydrogen sulfide adsorption in the MIL-53 (Al, Cr, Fe), MIL-47 (V), MIL-100 (Cr), and MIL-101 (Cr) metal-rganic frameworks at room temperature. *J. Am. Chem. Soc.* **2009**, *131*, 8775–8777. [CrossRef]
22. Chen, L.; Mowat, J.P.S.; Fairen-Jimenez, D.; Morrison, C.A.; Thompson, S.P.; Wright, P.A.; Düren, T. Elucidating the breathing of the metal–organic framework MIL-53 (Sc) with ab initio molecular dynamics simulations and in situ X-ray powder diffraction experiments. *J. Am. Chem. Soc.* **2013**, *135*, 15763–15773. [CrossRef] [PubMed]
23. Boutin, A.; Coudert, F.X.; Springuel-Huet, M.A.; Neimark, A.V.; Férey, G.; Fuchs, A.H. The behavior of flexible MIL-53 (Al) upon CH4 and CO_2 adsorption. *J. Phys. Chem. C* **2010**, *114*, 22237–22244. [CrossRef]
24. Liu, Y.; Her, J.-H.; Dailly, A.; Ramirez-Cuesta, A.J.; Neumann, D.A.; Brown, C.M. Reversible structural transition in MIL-53 with large temperature hysteresis. *J. Am. Chem. Soc.* **2008**, *130*, 11813–11818. [CrossRef] [PubMed]
25. Beurroies, I.; Boulhout, M.; Llewellyn, P.L.; Kuchta, B.; Férey, G.; Serre, C.; Denoyel, R. Using pressure to provoke the structural transition of metal-rganic frameworks. *Angew. Chem. Int. Ed.* **2010**, *49*, 7526–7529. [CrossRef]
26. Neimark, A.V.; Coudert, F.-X.; Triguero, C.; Boutin, A.; Fuchs, A.H.; Beurroies, I.; Denoyel, R. Structural transitions in MIL-53 (Cr): View from outside and inside. *Langmuir* **2011**, *27*, 4734–4741. [CrossRef]
27. Serre, C.; Bourrelly, S.; Vimont, A.; Ramsahye, N.A.; Maurin, G.; Llewellyn, P.L.; Daturi, M.; Filinchuk, Y.; Leynaud, O.; Barnes, P.; et al. An explanation for the very large breathing effect of a metal–organic framework during CO_2 adsorption. *Adv. Mater.* **2007**, *19*, 2246–2251. [CrossRef]
28. Mounfield, W.P., III; Walton, K.S. Effect of synthesis solvent on the breathing behavior of MIL-53 (Al). *J. Colloid Interface Sci.* **2015**, *447*, 33–39. [CrossRef]
29. Heymans, N.; Vaesen, S.; De Weireld, G. A complete procedure for acidic gas separation by adsorption on MIL-53 (Al). *Microporous Mesoporous Mater.* **2012**, *154*, 93–99. [CrossRef]
30. Deniz, E.; Karadas, F.; Patel, H.A.; Aparicio, S.; Yavuz, C.T.; Atilhan, M. A combined computational and experimental study of high pressure and supercritical CO_2 adsorption on Basolite MOFs. *Microporous Mesoporous Mater.* **2013**, *175*, 34–42. [CrossRef]
31. Camacho, B.C.R.; Ribeiro, R.P.P.L.; Esteves, I.A.A.C.; Mota, J.P.B. Adsorption equilibrium of carbon dioxide and nitrogen on the MIL-53 (Al) metal organic framework. *Sep. Purif. Technol.* **2015**, *141*, 150–159. [CrossRef]
32. Gonzalez-Nelson, A.; Coudert, F.X.; van der Veen, M.A. Rotational dynamics of linkers in metal-organic frameworks. *Nanomaterials* **2019**, *9*, 330. [CrossRef]
33. Hefti, M.; Joss, L.; Bjelobrk, Z.; Mazzotti, M. On the potential of phase-change adsorbents for CO_2 capture by temperature swing adsorption. *Faraday Discuss.* **2016**, *192*, 153–179. [CrossRef]
34. Liu, B.; Li, Y.; Hou, L.; Yang, G.; Wang, Y.Y.; Shi, Q.-Z. Dynamic Zn-based metal-organic framework: Step wise adsorption, hysteretic desorption and selective carbon dioxide uptake. *J. Mater. Chem.* **2013**, *1*, 6535–6538. [CrossRef]
35. Ribeiro, R.P.P.L.; Barreto, J.; GrossoXavier, M.D.; Martins, D.; Esteves, I.A.A.C.; Branco, M.; Tirolien, T.; Mota, J.P.B.; Bonfait, G. Cryogenic neon adsorption on Co3 (ndc) 3 (dabco) metal-organic framework. *Microporousand Mesoporous Mater.* **2020**, *298*, 110055. [CrossRef]
36. Lyubchyk, A.; Esteves, I.A.A.C.; Cruz, F.J.A.L.; Mota, J.P.B. Experimental and theoretical studies of supercritical methane adsorption in the MIL-53 (Al) metal organic framework. *J. Phys. Chem.* **2011**, *115*, 20628–20638. [CrossRef]
37. Ribeiro, R.P.P.L.; Camacho, B.C.R.; Lyubchyk, A.; Esteves, I.A.A.C.; Cruz, F.J.A.L.; Mota, J.P.B. Experimental and computational study of ethane and ethylene adsorption in the MIL-53 (Al) metal organic framework. *Microporousand Mesoporous Mater.* **2016**, *230*, 154–165. [CrossRef]
38. Dreisbach, F.; Staudt, R.; Keller, J.U. High pressure adsorption data of methane, nitrogen, carbon dioxide and their binary and ternary mixtures on activated carbon. *Adsorption* **1999**, *5*, 215–227. [CrossRef]
39. Coudert, F.-X.; Jeffroy, M.; Fuchs, A.H.; Boutin, A.; Mellot-Draznieks, C. Thermodynamics of guest-induced structural transitions in hybrid organic−inorganic frameworks. *J. Am. Chem. Soc.* **2008**, *130*, 14294–14302. [CrossRef] [PubMed]
40. Mota, J.P.B.; Martins, D.; Lopes, D.; Catarino, I.; Bonfait, G. Structural transitions in the MIL-53 (Al) metal–organic framework upon cryogenic hydrogen adsorption. *J. Phys. Chem.* **2017**, *121*, 24252–24263. [CrossRef]
41. Hiraide, S.; Sakanaka, Y.; Kajiro, H.; Kawaguchi, S.; Miyahara, M.T.; Tanaka, H. High-through put gas separation by flexible metal–organic frameworks with fast gating and thermal management capabilities. *Nat. Commun.* **2020**, *11*, 3867. [CrossRef]
42. Ghysels, A.; Vanduyfhuys, L.; Vandichel, M.; Waroquier, M.; Van Speybroeck, V.; Smit, B. On the thermodynamics of framework breathing: A free energy model for gas adsorption in MIL-53. *J. Phys. Chem.* **2013**, *117*, 11540–11554. [CrossRef]
43. Jeffroy, M.; Fuchs, A.H.; Boutin, A. Structural changes in nanoporous solids due to fluid adsorption: Thermodynamic analysis and Monte Carlo simulations. *Chem. Commun.* **2008**, 3275–3277. [CrossRef] [PubMed]
44. Ferreira, T.J.; Vera, A.T.; De Moura, B.A.; Esteves, L.M.; Tariq, M.; Esperança, J.M.S.S.; Esteves, I.A.A.C. Paramagnetic ionic liquid/metal organic framework composites for CO_2/CH4 and CO_2/N2 separations. *Front. Chem.* **2020**, *8*, 590191. [CrossRef] [PubMed]

45. Nabais, A.R.; Ribeiro, R.P.P.L.; Mota, J.P.B.; Alves, V.D.; Esteves, I.A.A.C.; Neves, L.A. CO_2/N_2 gas separation using Fe (BTC)-based mixed matrix membranes: A view on the adsorptive and filler properties of metal-organic frameworks. *Sep. Purif. Technol.* **2018**, *202*, 174–184. [CrossRef]
46. Poling, B.E.; Prausnitz, J.M.; O'Connell, J.P. *The Properties of Gases and Liquids*; McGraw-Hill: New York, NY, USA, 2001.

energies

Article

Extrusion and Characterization of High Si/Al Ratio ZSM-5 Using Silica Binder

Ana Almeida [1], Rui P. P. L. Ribeiro [2], José P. B. Mota [2] and Carlos Grande [1,*]

[1] SINTEF Industry, Forskningsveien 1, 0373 Oslo, Norway; Ana.Almeida@theOGTC.com

[2] LAQV-REQUIMTE, Department of Chemistry, Faculty of Science and Technology, Universidade Nova de Lisboa, 2829-516 Caparica, Portugal; rpp.ribeiro@fct.unl.pt (R.P.P.L.R.); pmota@fct.unl.pt (J.P.B.M.)

* Correspondence: carlos.grande@sintef.no; Tel.: +47-93207532

Received: 7 February 2020; Accepted: 3 March 2020; Published: 5 March 2020

Abstract: Biogas upgrading is a key operation for transforming raw biogas into valuable biomethane that can be used as fuel or transported through pipelines. Pressure swing adsorption (PSA) is one possible technique that can be used for upgrading. ZSM-5 with high silica/aluminum (Si/Al) ratio has a reasonable CO_2/CH_4 selectivity and an almost linear CO_2 adsorption isotherm, which can reduce power consumption. Extrusion of zeolites uses Al-based binders which can result in a denaturation and in a decrease of Si/Al ratio, promoting a steeper CO_2 isotherm and also impacting the water adsorption. In this work, we have extruded a ZSM-5 (with a Si/Al = 200) using only silica-based binder. Different samples were obtained using different extrusion paste compositions and operating conditions and their textural properties characterized. The mechanical strength of the samples as well as the CO_2, CH_4, and H_2O adsorption equilibrium isotherms at 303–343 K were measured. Our results show that it is possible to produce extrudates with mechanical resistance comparable to (or higher than) commercial zeolite materials with surface area reductions lower than 10% and little or no impact on the CO_2/CH_4 selectivity.

Keywords: zeolite; ZSM-5; adsorption; biogas upgrading; shaping; extrusion

1. Introduction

Over the last 50 years, there has been an enormous expansion of the available porous materials. The vast possibilities of surface customization have boosted the research on separation technologies, like membranes and adsorption-based processes [1,2]. However, for new materials, there is a large gap between discovery and its utilization in industrial applications. Such gap lays primarily in having a good knowledge of proper practices for the shaping of powders into millimeter-scale particles [3–7]. The challenge is to have particles with adequate hardness and size to be used in large columns but with minimal changes to their properties as powders. For adsorption processes, particles are in the mm-scale to keep the pressure drop controlled while limiting diffusion effects. Novel applications in moving-bed type of reactors require sub-mm particle diameters [8]. The typical shapes of industrial porous materials are pellets, extrudates, laminates, and monoliths [9].

Several methods can be used for shaping porous materials. The most industrialized technologies are extrusion [4,10–15] and spray drying [16–19]. Many other technologies are at the research and demonstration stages, and the details are not published in many cases since this is the pre-commercial step for a material. Some of the technologies that can find a commercial path in the near future are hydrocolloids [20,21], powder compression [22–24], and casting [25]. Recently, 3D printing is being used for the production of shapes [26–29].

Due to its fast production rate, extrusion is nowadays the technology most employed for shaping porous materials. The term extrusion describes a process involving the movement of a material through a confined space using little-to-no solvent. Extrusion started as a technology for shaping metals into pipes and has been extended into polymers, pharmaceuticals, food, etc. In extrusion, the movement of the material can be propelled by a hydraulic medium (piston) or by screws. In order to be extruded, the solid material should have appropriate plastic properties. Since the shear forces in extrusion can be intense, a material without the correct viscosity, for example, can be heated to extreme temperatures, possibly damaging its structure or completely changing its properties.

Solid porous materials are normally mixed into pastes composed of a "binder" (or mixture of binders) that provides mechanical strength and moldability, and a plasticizer that serves as wetting agent and decreases the viscosity of the mixture, forming a plastic composition. Optionally, a dispersant can be added to avoid agglomeration [30–32]. Finding the right combination of the correct ingredients that will produce a successful paste is challenging, and there is currently no theoretical approach for this.

ZSM-5 is a well-known commercial zeolite. It is possible to control its acidity to a great extent by controlling the Si/Al ratio. It is known that the adsorption properties of ZSM-5 (and also other zeolites) depend on its Si/Al ratio [33–35]. When shaping zeolites, clays, alumina, and silica are commonly used to provide hardness to the extrudates. Organic additives may also be added, increasing the strength of the extrudate and reducing shape variations during extrusion. Cellulose, methylcellulose, and polyvinyl alcohol (PVA) are the most common organic additives. They also contribute to generate macroporosity in the final shaped particles. Water is the most used plasticizer. After the paste is extruded, it is normally dried to remove the plasticizer and then "fired" to ~750 K where some chemical reactions (or sintering) take place: methyl siloxane ether transforms into silica [36] and kaolin into metakaolin [32]. The organic additives are burned generating macroporosity.

In the case of using a ZSM-5 with high Si/Al ratio, it is essential that the elements composing the paste do not result in a change of adsorption properties. Carbon dioxide and water adsorption isotherms are very sensible to changes in Si/Al ratio and cation substitutions, due to their quadrupole moment and polarity, respectively. In adsorption processes, the adsorption "strength" yielded by the initial slope of the isotherm can result in higher capacity, but it also results in the need for regenerating at a lower pressure increasing power consumption.

In this work, we have extruded high silica ZSM-5 (Si/Al = 200) with pure silica binders aiming to avoid acidification of the zeolite during the extrusion process. The prepared samples also used an organic binder (PVA) and water as plasticizer. Different paste compositions were used for extrusion, and the resulting samples were characterized in terms of surface area and mechanical strength. The most successful sample was used for measurement of adsorption equilibrium of CH_4 and CO_2 and H_2O at 303, 323, and 343 K.

2. Materials and Methods

2.1. Chemicals

A commercial sample of zeolite ZSM-5 (Clariant, Muttenz, Switzerland) with a Si/Al ratio of 200 was employed as the target porous solid to be shaped. Polyvinyl alcohol (PVA, Aldrich, Schnelldorf, Germany) and colloidal silica (LUDOX®LS, 220 m^2/g, 30 wt %, Aldrich, Schnelldorf, Germany) were used as temporary and permanent binders, respectively. Deionized water was employed as plasticizer.

Pure gases employed in the adsorption equilibria measurements were provided by Yara (Oslo, Norway): CH_4 (purity >99.995%) and CO_2 (purity >99.9992%).

2.2. ZSM-5 Extrudates Preparation

ZSM-5 extrudates were made in a Thermo Scientific (Karlsruhe, Germany) HAAKE MiniLab II twin-screw extruder equipped with two co-rotating screws and a 2 mm diameter dye. The extruder is equipped with a backflow channel and an automatic bypass valve that allow the operator to work with well-defined residence times. The extruder can directly push the sample from the feeder to the dye in a normal extrusion process that we denominate here as "open cycle". Another option that the extruder has is that it allows to recirculate the same paste from the end of the extruder to the feed point to obtain better mixing. Operation under the recycling mode is termed here as "closed loop". Extrusion in open and closed cycles were tested. Preliminary experiments with long recirculation times indicated that similar results are obtained for recirculation times over 5 minutes. For this reason, in this work, the extruder was operated with a recirculation time of 5 minutes emulating longer processes.

For extrusion purposes, pastes were prepared containing powdered ZSM-5 zeolite, PVA (temporary binder), colloidal silica (permanent binder), and water as plasticizer. The pastes were prepared by mixing ZSM-5 and PVA powders followed by addition of water and LUDOX®in appropriate amounts. Several paste compositions were prepared and extruded to evaluate the influence of the zeolite/binder ratio on the paste properties and obtained extrudates. The composition of the samples successfully extruded are detailed on Table 1.

Following extrusion, the samples were air-dried at constant temperature (298 K) for at least 24 hours. Subsequently, the extrudates underwent a firing stage, which consisted of gradually heating the samples to 773 K for 12 hours to remove the plasticizer and temporary binder, generating macroporosity within the extruded sample.

2.3. *Characterization of ZSM-5 Extrudates*

The samples of ZSM-5 extrudates prepared were thoroughly characterized. All samples underwent compression tests, scanning electron microscopy (SEM), and N_2 adsorption at 77 K for surface area assessment. Furthermore, the best sample was selected for measurement of CH_4, CO_2, and H_2O adsorption equilibrium isotherms up to 1 bar at 303, 323, and 343 K.

The mechanical stability and stiffness of the prepared extrudates were evaluated through compression/crushing tests. For this purpose, a Zwick/Roell Z250 universal test machine equipped with a 500 N load cell was employed. For each assay, a sample was placed on the center of the lower compression plate, which was lifted against the upper plate at a rate of 0.5 mm/min. The assay was performed three times for each sample using a single pellet each time (diameter ~2 mm, length ~5 mm). TestXpert II software was used to record the stress–strain curve, i.e., applied force (N) as a function of the sample deformation (mm).

Scanning electron microscopy (SEM) was performed in a Zeiss Auriga CrossBeam FIB/SEM workstation available at Cenimat/i3N (FCT-NOVA). SEM images were taken using 5.0 kV acceleration voltage.

N_2 adsorption at 77 K was determined in a BELSORP-mini II instrument. The ZSM-5 extrudates samples were externally degassed at 423 K overnight followed by 2 hours degassing in the unit at 423 K under vacuum. The obtained data were employed to determine the BET surface area and specific pore volume of the samples.

2.4. *Adsorption Equilibrium of CO_2, CH_4, and H_2O*

CO_2, CH_4, and H_2O pure component adsorption equilibrium isotherms were determined at 303, 323, and 343 K. Data were collected using a Belsorp-MAX volumetric apparatus: in the 0–1 bar pressure range for CH_4 and CO_2 and up to 4 kPa for water vapor. The degassing procedure was the same employed prior to the N_2 isotherms. Both adsorption and desorption data were measured to confirm the reversibility of the process and check for hysteresis.

3. Results

3.1. *Characterization of ZSM-5 Extrudates*

In this work, 21 samples were prepared in order to study the effect of the different components in the production of high Si/Al ZSM-5 extrudates. The paste preparation showed that the content of additives greatly affects the consistency of the extrusion paste and its ability to undergo the extrusion process. Some of the pastes could not be extruded at all, and some others were not characterized because their shape indicated that the extrusion was not successful. In this publication, only the successfully extruded ZSM-5 are presented. Extrusion using a wide range of paste compositions consequently resulted in extrudates with different features. Figure 1 shows the binder composition of the ZSM-5 extrusion pastes and the effect of this composition on extrudate shape and appearance.

The particle density (ρ_p) of each extruded sample was calculated by measurement of the mass (w), length (l), and diameter (d) of each pellet. To reduce error in the measurements, the corners were flattened with a scalpel. Based on the measurements, the volume of each pellet ($V = \pi\frac{d^2}{4} \times l$) was used to calculate the particle density via $\rho_p = \frac{w}{V}$.

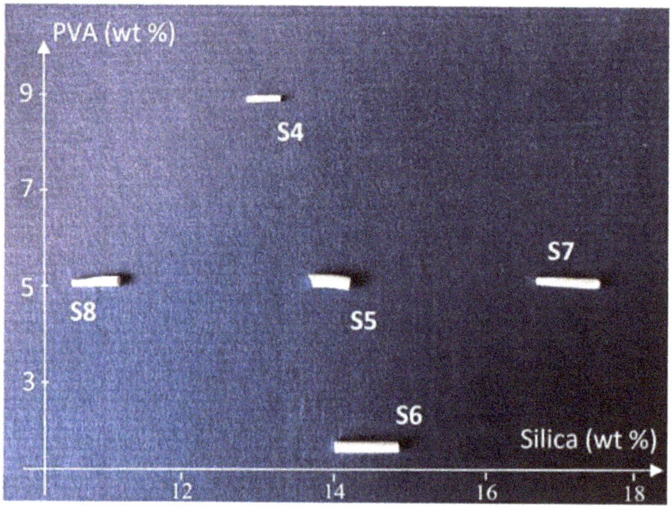

Figure 1. Effect of binder composition on the morphology of ZSM-5 extrudates.

Table 1 summarizes the density results obtained on the effect of paste composition. Sample 4 has high PVA content, and the extruded paste was very viscous and flexible. After thermal treatment, this resulted in uneven extrudates with irregular diameters and low mechanical resistance. Extrusion in closed cycles slightly improved consistency, however, these samples were not selected for further characterization. Sample 5, with intermediate PVA content, resulted in consistent extrudates. Extrusion in closed cycles produced stronger extrudates than in open cycles with shorter extrusion times. Sample 6 has low PVA content and produced extrudates with regular shape and diameter. A sample with the same composition was tested under closed cycles and the results were not satisfactory.

Table 1. Composition of ZSM-5-based extrusion pastes and extrudates density.

Sample	Preparation	ZSM-5 (wt %)	PVA (wt %)	Silica (wt %)	Density (g/cm^3)
4	Open cycle	77.4	9.0	13.6	N.A.
5	Closed cycle	81.0	5.0	14.0	1.17
6	Open cycle	83.7	2.0	14.3	1.03
7	Closed cycle	78.0	5.0	17.0	0.98
8	Closed cycle	84.0	5.0	11.0	1.14
16	Closed cycle	80.0	5.0	15.0	1.13

Samples 7 and 16, with high silica content, resulted in extrudates with a very irregular surface and slight diameter variations. Closed-cycle extrusion resulted in higher viscosity and better surface uniformity. A density of 1.13 g/cm^3 was obtained with 15% silica content in the paste and 0.98 g/cm^3

with 17%, indicating that the increase over a certain level does not improve the properties of the extrudate. Sample 8 has low silica content and resulted in regularly shaped extrudates. Extrusion in closed cycles improved surface smoothness and increased viscosity causing greater diameter variations.

In order to select samples for further testing, there are two main parameters that should be observed: the density and the surface rugosity. We have highlighted the importance of the density in the sample preparation because if the shaping technique produces samples with low densities, larger columns will be required to pack a similar weight of material. Moreover, if the samples present a very irregular surface, when they are packed, they will generate dust that can increase the pressure drop along the columns and may ultimately damage other equipment like valves, etc.

3.1.1. Compression Tests

The mechanical stability of the extrudates was assessed by subjecting the samples to compression/crushing tests. In these assays, a piston applies a force that increases as the sample is compressed. At the precise moment that the extrudate breaks, there is a sudden drop in the exerted force. The value recorded prior to this decrease is reported as the crushing force or crushing strength of the material. Figure 2 shows a comparison of the results obtained for two extrudates of the same batch and, also, the crushing force required to break an extrudate of commercial 4A zeolite [9]. The obtained results clearly show that despite the variability between pellets of the same batch, the extruded ZSM-5 pellets have good mechanical properties comparable to (or even higher than) other commercially available zeolites.

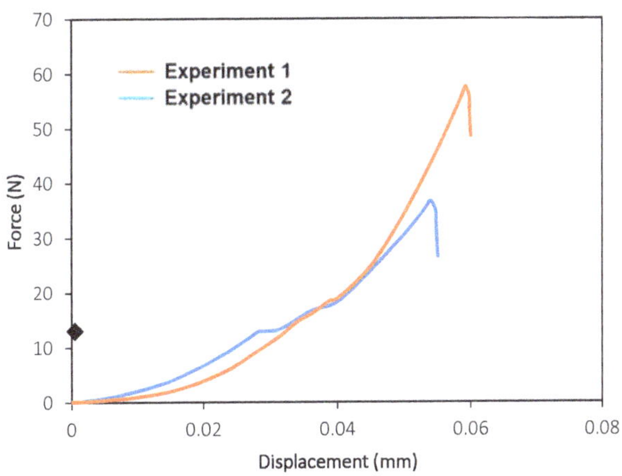

Figure 2. Results of the compression test for two pellets of the same ZSM-5 extrudate sample. (The solid symbol represents the crushing force for a commercial 4A zeolite extrudate, reported for comparison [9]).

Figure 3 shows the averaged crushing force as a function of the silica content in the extruded paste (PVA is removed during the post-extrusion thermal treatment). The reported average was taken over three different assays performed for each sample. The results suggest that a correlation between silica content and hardness in ZSM-5 extrudates exists, although higher amounts of silica are detrimental to the mechanical resistance of the material. This may be explained by the irregular surfaces and diameter variations observed in samples with higher silica content. In those cases (higher amount of silica), the viscosity of the paste increases resulting in a higher pressure inside the extruder. When the paste exits through the die, there is a sudden decompression that results in a solid expansion and in a rough surface. Such decompression clearly affects the mechanical properties.

3.1.2. Scanning Electron Microscopy (SEM) Imaging

The morphology of the ZSM-5 extrudates and the starting ZSM-5 powder were analyzed by SEM imaging. Figure 4a–c show images of the surfaces of the extrudates with their average diameters and without visible cracks at the surface. Some surface rugosity can be observed in sample 7 and, to a lesser extent, in sample 8. Sample 5 has a very smooth surface, which is a very positive aspect of this sample. SEM imaging showed that although the same die, with 2 mm diameter, was used in all the extrusions, the final diameter of the extrudates were variable for each batch. Figure 4a–c show that the extrudates can present diameters of 2.0 mm (sample 5), 1.76 mm (sample 8), and 1.31 mm (sample 7). In Figure 4c, it is possible to observe a flat area in one of the sides of the ZSM-5 extrudate. This occurred as the material was placed on a flat surface to dry following extrusion. The materials that deform after extrusion and present an extensive degree of shrinkage are not optimal in their composition. Figure 4d–f presents the SEM images obtained for the extrudates (samples 7 and 8 and the zeolite powder; the SEM images demonstrate that the material underwent extrusion without suffering deformation. The crystal morphology remains apparently unchanged, although a tighter crystal packing and the predominance of smaller and more homogenous structures in the extrudate are observed.

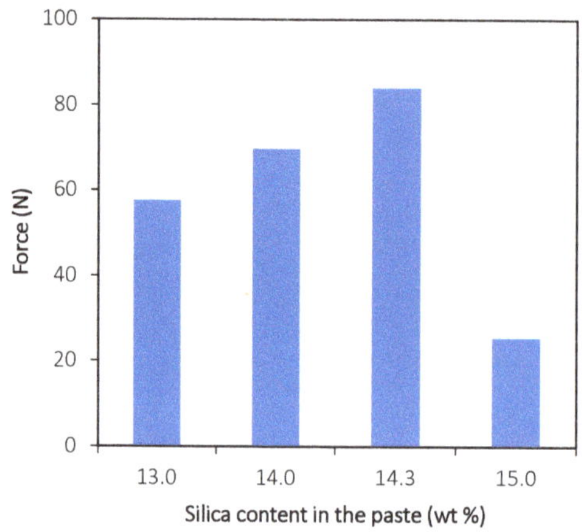

Figure 3. Maximum force applied on ZSM-5 extrudates versus silica content in the extruded paste.

3.1.3. Nitrogen Adsorption at 77 K

The N_2 adsorption isotherms obtained at 77 K showed type I behavior indicative of microporous samples with pore filling at high relative pressures. The N_2 adsorption isotherms obtained at 77 K for the pristine powder and extrudates in sample 5 are shown in Figure 5a. The BET surface area was determined from the experimental data and the results obtained for the various silica content on the extrudates after PVA removal through a firing procedure is displayed in Figure 6.

Table 2 presents a comparison between the textural results obtained for sample 5 and the values measured in the ZSM-5 powder. The BET surface area of the zeolite extrudates decreased between 7% and 11% in relation to the original ZSM-5 BET surface area. The observation of a decrease in the available surface area was expected due to the presence of silica binder in the extrudates. Despite some dispersion in the results, there seems to be an approximately linear correlation between the surface area and the silica content in the extrudates, as shown in Figure 6. Assuming that all PVA content is removed from the final extrudates and that SiO_2 has a surface area of 220 m^2/g, the theoretical BET that should be obtained for a solid adsorbent composed by ZSM-5 and SiO_2, if no detrimental effects in the surface/pore occur, can be calculated. This theoretical line is also shown in Figure 6 and compared with the experimental values obtained as a function of the silica content in the extrudates (assuming no remaining PVA). It can be seen that the obtained results are only slightly lower than the expected results when considering the surface area originally reported for the SiO_2 (LUDOX®) employed in this study.

Figure 4. SEM images of the ZSM-5 extrudates at 40× magnification: (**a**) sample 5, (**b**) sample 7, and (**c**) sample 8. SEM images of: (**d**) the ZSM-5 powder and extrudates of (**e**) samples 7 and (**f**) 8 at 20,000× magnification.

Table 2. Textural results for best formulation compositions.

ZSM-5 Sample	A_{BET} (m^2/g)	Micropore Volume (cm^3/g)
Powder	341	0.305
Extrudates (Sample 5)	318	0.260

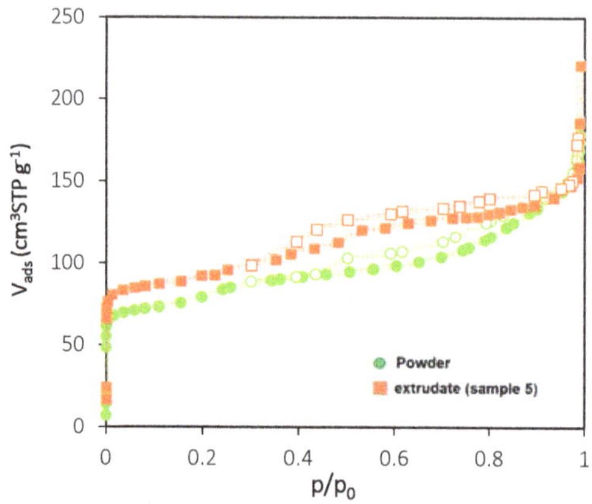

Figure 5. N_2 adsorption isotherms at 77 K for ZSM-5 pristine powder (■) and extrudate sample 5 (•). The filled and empty symbols denote adsorption and desorption data, respectively.

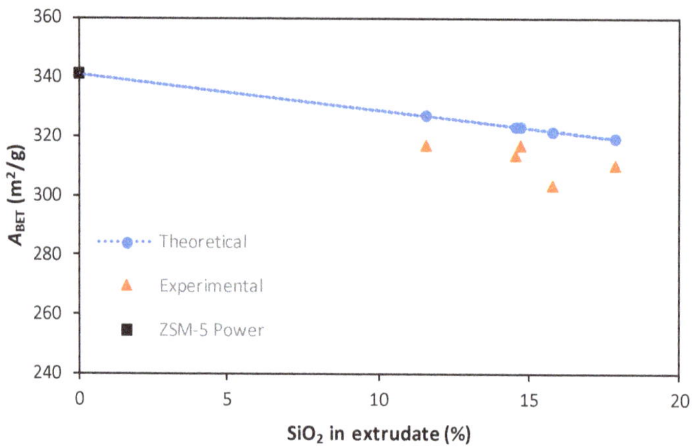

Figure 6. BET surface area as a function of the silica content in the extrudate after thermal treatment at 773 K. The experimental data are compared with the theoretical value considering the original BET surface areas from the ZSM-5 and the amount of SiO_2 employed.

From the compression tests, SEM imaging, and N_2 adsorption experiments, we can conclude that an intermediate amount of silica (14% as used to produce sample 5) resulted in a very good quality sample with less surface rugosity and higher density. For this reason, sample 5 was used for adsorption characterization.

3.2. *Adsorption Equilibrium of CH_4, CO_2, and H_2O*

Figures 7 and 8 present the adsorption equilibria of CH_4 and CO_2 on ZSM-5 extrudates sample 5 at 303, 323, and 343 K up to 100 kPa. The shape of the isotherms is Type I, and the results show that the CH_4 experimental data were successfully fitted with the Langmuir model; in the case of CO_2 adsorption equilibrium, the Langmuir fitting is rather poor, especially in the low-pressure region. In this case, the dual-site Langmuir (DSL) model presents very good fitting of the experimental data. The Langmuir and DSL fitting parameters are listed in Table 3 along with the model details. The results obtained for CH_4 adsorption on ZSM-5 extrudates is higher than reported for Na-ZSM-5 with Si/Al = 130 [37]. For example, at 100 kPa, the CH_4 adsorption capacity of our sample is more than double (~2.3 times) when compared with the lower Si/Al Na-ZSM-5 sample. However, the results are comparable with other studies made with higher Si/Al ratios, where the differences observed at 100 kPa are smaller than 4% [38,39]. The amount of CO_2 is also comparable with previous data of ZSM-5 with higher Si/Al ratio, which is lower than in the case of more acidic samples with other cations. For example, at 90 kPa and 303 K, the amount of adsorbed CO_2 shows a difference as low as 2% when compared with the data previously reported by Heymans et al. [39]. On the other hand, our sample shows a decrease in CO_2 adsorption capacity of approximately 18% and 30% when compared with samples with lower Si/Al ratios of 130 [37] and 30 [40], respectively. Despite lower selectivity of our sample, the shape of the isotherms are more linear and, thus, the regeneration in a PSA process can be performed at a higher pressure, reducing the overall power consumption.

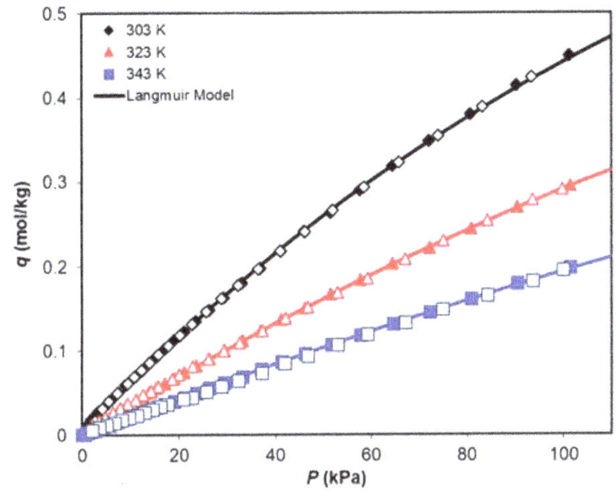

Figure 7. Adsorption equilibrium isotherms of CH_4 on ZSM-5 extrudates at 303, 323, and 343 K. The filled symbols denote adsorption data and open symbols denote desorption data. Solid lines represent Langmuir model fitting.

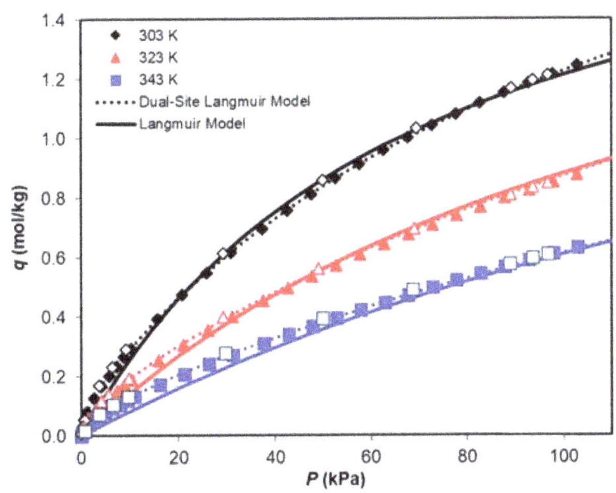

Figure 8. Adsorption equilibrium isotherms of CO_2 on ZSM-5 extrudates at 303, 323, and 343 K. The filled symbols denote adsorption data and open symbols denote desorption data. Solid and dashed lines represent the fittings obtained using the Langmuir and dual-site Langmuir models, respectively.

Table 3. Langmuir and dual-site Langmuir fitting parameters for CH_4, CO_2, and H_2O adsorption on ZSM-5 extrudates.

	Langmuir Model					
	$q_i = \frac{q_{max,i} K_i P}{1+K_i P}$			$K_i = K_i^0 \exp\left(\frac{-\Delta H}{RT}\right)$		
	q_{max} (mol/kg)			$K_A{}^0$ (kPa^{-1})		$-\Delta H$ (J/mol)
CO_2	2.04			3.79×10^{-7}		26,600
CH_4	1.47			5.70×10^{-7}		22,500
	Dual-Site Langmuir Model					
	$q_i = \frac{q_{max1,i} K_{1,i} P}{1+K_{1,i} P} + \frac{q_{max2,i} K_{2,i} P}{1+K_{2,i} P}$			$K_i = K_i^0 \exp\left(\frac{-\Delta H}{RT}\right)$		
	q_{max1} (mol/kg)	$K_1{}^0$ (kPa^{-1})	$-\Delta H_1$ (J/mol)	q_{max2} (mol/kg)	$K_2{}^0$ (kPa^{-1})	$-\Delta H_2$ (J/mol)
CO_2	2.360	3.77×10^{-7}	25,417	0.106	1.39×10^{-7}	39,886
H_2O	0.869	2.55×10^{-8}	45,071	0.187	7.91×10^{-5}	37,394

Figure 9 shows the isosteric heats of adsorption (Q_{st}) for CO_2 and CH_4 as a function of the respective loadings. Q_{st} has been determined directly from the experimental adsorption equilibria data, as represented in Figure 9 by the empty (CH_4) and solid symbols (CO_2), by fitting it to 6th-order polynomials and applying the integrated form of the Clapeyron equation. The study of isosteric heat confirms the occurrence of physisorption for both CO_2 and CH_4 (Q_{st} < 80 kJ/mol). CO_2 adsorption involves higher interaction energy than CH_4, although both components show variation of Q_{st} with loading, indicative of energetically heterogenous behavior. The results presented in Figure 9 show that the ΔH parameter employed in the Langmuir fitting (solid lines), reported in Table 3, is in good agreement with the calculated values of Q_{st}.

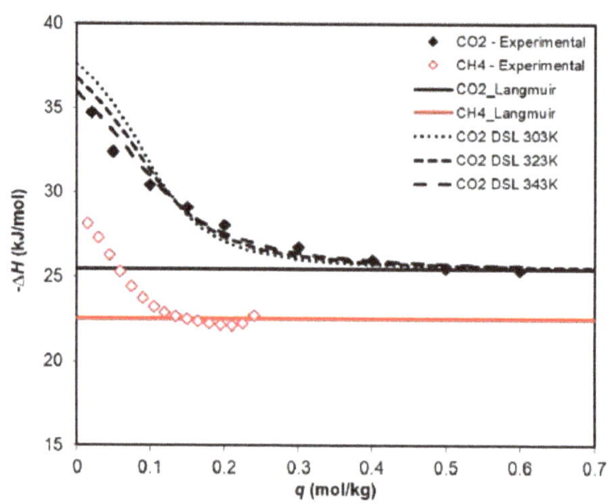

Figure 9. Isosteric heat as a function of loading for CO_2 and CH_4 on ZSM-5. The closed and open symbols denote the isosteric heat calculated from the isosteric data derived from the experimental adsorption isotherms for CO_2 and CH_4, respectively. The solid lines represent the isosteric heat obtained from the Langmuir fitting; the dashed lines represent the values obtained with the DSL for CO_2.

The pure-gas isosteric heat derived from the DSL model was determined by applying the following expression [41]:

$$Q_{st} = \frac{\Delta H_1 q_{sat,1} K_1 (1+K_2 P)^2 + \Delta H_2 q_{sat,2} K_2 (1+K_1 P)^2}{q_{sat,1} K_1 (1+K_2 P)^2 + q_{sat,2} K_2 (1+K_1 P)^2}.$$

The heterogeneity consideration in the DSL model allows the obtained isosteric heat to describe, very well, the values obtained using the polynomial fitting and integrated form of the Clapeyron equation. This is observed by the good agreement between the dashed lines and the filled symbols in Figure 9. Furthermore, it is observed that Q_{st} is mostly independent of temperature within the studied temperature range.

Figure 10 presents the CO_2/CH_4 ideal equilibrium selectivity, calculated based on data from pure gases. The study of selectivity, made by comparison of adsorption capacities of CO_2 and CH_4 under the same pressure and temperature, shows that the extrudates hold higher affinity to CO_2 than CH_4. This feature is highly pressure-dependent, decreasing significantly as the pressure increases.

Figure 11 presents the adsorption equilibria of water vapor on ZSM-5 extrudates at 303, 323, and 343 K. The solid lines correspond to the fitting of the DSL model using the parameters reported in Table 3. The fitting with the Langmuir model was not very accurate. Previous studies report that the saturation concentration of water in ZSM-5 presents significant variation in connection with Si/Al ratios and the presence of defects within the porous material [42,43]. For this reason, due to the high Si/Al ratio in the ZSM-5 extrudates, a significantly lower H_2O uptake was obtained when compared with previously reported data [37]. For example, at 323 K and 1.8 kPa, our sample adsorbs less than half the amount of H_2O than for the referred lower Si/Al ratio sample. This is another advantage of having an adsorbent with high Si/Al ratio where the co-adsorbed water will have a much smaller influence on the CO_2/CH_4 ratio and loading amounts. As an example, we present in Figure 12 the prediction of the CO_2 loading in the presence of 0.2 and 1.0 kPa of water at 303 K using the multicomponent extension of the DSL model. The loading of CO_2 is reduced, but not as drastically as in other zeolites.

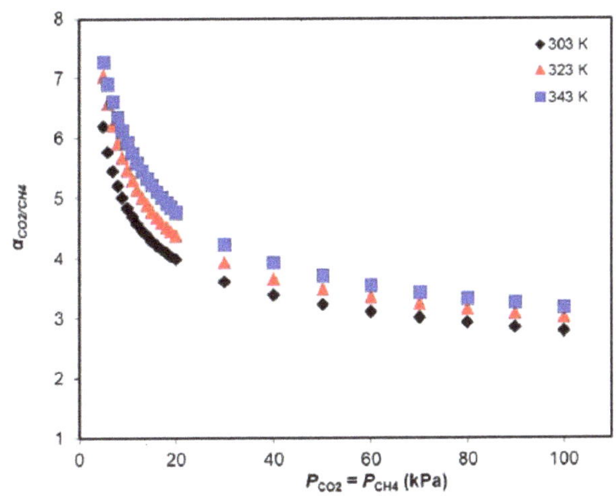

Figure 10. CO_2/CH_4 ideal selectivity as a function of the partial pressure in an equimolar mixture at 303 K (♦), 323 K (▲), and 343 K (■).

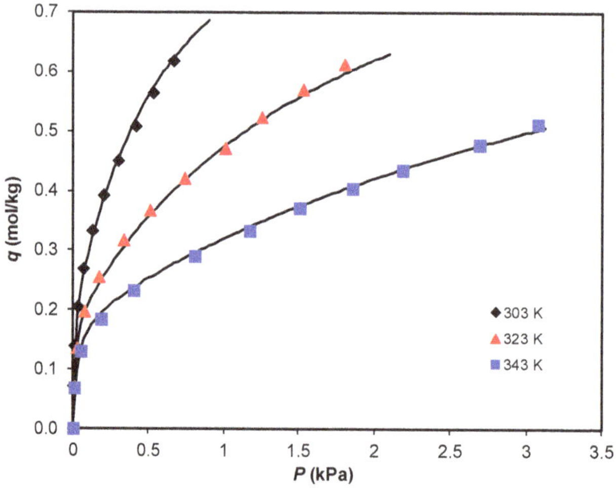

Figure 11. Adsorption equilibrium isotherms of H_2O on ZSM-5 extrudates at 303 K (♦), 323 K (▲), and 343 K (■). Solid lines represent the dual-site Langmuir model fittings.

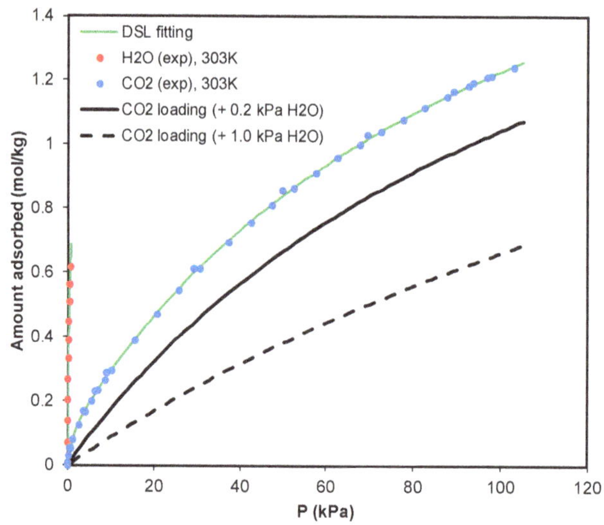

Figure 12. Adsorption equilibrium isotherms of H_2O and CO_2 on ZSM-5 extrudates at 303 K and the prediction of multicomponent CO_2 loading using 0.2 and 1.0 kPa partial pressure of water.

4. Conclusions

This study has focused on the extrusion of zeolite ZSM-5 with high Si/Al ratios using silica binders to avoid zeolite denaturation during shaping. The obtained extrudates were characterized by measuring their density, surface area, scanning electron microscopy, and crushing strength. The best sample (sample 5) was selected for measurement of adsorption equilibrium of CH_4, CO_2, and H_2O to evaluate the viability of the application of these materials for biogas upgrading.

Experiments were made with different amounts of silica as permanent binder and of polyvinyl alcohol (PVA) as a macroporosity generator. The optimal PVA content is approximately 5% while an amount of silica around 14% results in extrudates with higher density, good mechanical properties, low surface roughness, and low reduction of surface area.

The best sample of produced extrudates was also characterized in terms of adsorption equilibrium of CO_2, CH_4, and H_2O. Its high silica ratio resulted in an apparent lower CO_2/CH_4 selectivity when compared with other data available in the literature. However, the adsorption isotherms were less steep, which is reflected in the lower heat of adsorption. Also important is the very small amount of water that is adsorbed by our sample. This means that the adsorbent can still be selective to CO_2 in the presence of a considerable amount of humidity in the gas to be treated.

The prepared extrudates maintain a high Si/Al and have surface area reduction as low as 7% and good mechanical properties. The linearity of their CO_2 adsorption isotherms makes the ZSM-5 extrudates with high Si/Al ratio an interesting material to be used in adsorption processes for biogas upgrading in which CO_2 can be desorbed without using very low vacuum.

Author Contributions: All authors contributed to the manuscript and have read and agreed to the published version of the manuscript.

Funding: We acknowledge the support of the Research Council of Norway through the CLIMIT program by the SINTERCAP project (233818). This publication has been produced with partial support from the BIGCCS Centre, performed under the Norwegian research program Centres for Environment-friendly Energy Research (FME).

Acknowledgments: The author acknowledges the following partners for their contributions: ConocoPhillips, Gassco, Shell, Statoil, TOTAL, GDF SUEZ and the Research Council of Norway (193816/S60). Rui P. P. L. Ribeiro acknowledges FCT/MCTES for financial support through grant SFRH/BPD/103533/2014 and the Norma Transitória DL57/2016 Program Contract. This work was partially supported by the Associate Laboratory for Green Chemistry - LAQV which is financed by national funds from FCT/MCTES (UID/QUI/50006/2019).

Conflicts of Interest: The authors declare no conflicts of interest. The funders had no role in the design of the study; in the collection, analyses, or interpretation of data; in the writing of the manuscript, or in the decision to publish the results.

References

1. Ruthven, D.M. *Principles of Adsorption and Adsorption Processes*; Wiley-Interscience: New York, NY, USA, 1984.
2. Yang, R.T. *Gas Separation by Adsorption Processes*; Butterworth Publishers: Boston, MA, USA, 1987.
3. Perego, C.; Villa, P. Catalyst preparation methods. *Catal. Today* **1997**, *34*, 281–305. [CrossRef]
4. Masala, A.; Vitillo, J.G.; Mondino, G.; Martra, G.; Blom, R.; Grande, C.A.; Bordiga, S. Conductive ZSM-5-Based Adsorbent for CO2 Capture: Active Phase vs Monolith. *Ind. Eng. Chem. Res.* **2017**, *56*, 8485–8498. [CrossRef]
5. Omojola, T.; Cherkasov, N.; Rebrov, E.V.; Lukyanov, D.B.; Perera, S.P. Zeolite minilith: A unique structured catalyst for the methanol to gasoline process. *Chem. Eng. Process.* **2018**, *131*, 137–143. [CrossRef]
6. Armor, J.N.; Farris, T.S. Simultaneous exchange and extrusion of metal exchanged zeolites. *Appl. Catal. A* **1994**, *114*, L187–L190. [CrossRef]
7. Schwarz, S.; Kojima, M.; O'Connor, C.T. Effect of stirring, extrusion and pelletisation on high pressure propene oligomerisation and xylene isomerisation over ZSM-5. *Appl. Catal.* **1991**, *68*, 81–96. [CrossRef]
8. Grande, C.A.; Kvamsdal, H.; Mondino, G.; Blom, R. Development of Moving Bed Temperature Swing Adsorption (MBTSA) Process for Post-combustion CO2 Capture: Initial Benchmarking in a NGCC Context. *Energy Procedia* **2017**, *114*, 2203–2210. [CrossRef]
9. Hou, J.; Sapnik, A.F.; Bennett, T.D. Metal-organic framework gels and monoliths. *Chem. Sci.* **2020**, *11*, 310–323. [CrossRef]
10. Kusgens, P.; Zgaverdea, A.; Fritz, H.G.; Siegle, S.; Kaskel, S. Metal-Organic Frameworks in Monolithic Structures. *J. Am. Ceram. Soc.* **2010**, *93*, 2476–2479. [CrossRef]
11. Grande, C.A.; Águeda, V.I.; Spjelkavik, A.; Blom, R. An efficient recipe for formulation of metal-organic Frameworks. *Chem. Eng. Sci.* **2015**, *124*, 154–158. [CrossRef]
12. Hong, W.Y.; Perera, S.P.; Burrows, A.D. Manufacturing of metal-organic framework monoliths and their application in CO2 adsorption. *Microporous Mesoporous Mater.* **2015**, *214*, 149–155. [CrossRef]
13. Wu, D.; Tang, M. Effects of process factors on extrusion of hierarchically porous ZSM-5 zeolite. *Powder Technol.* **2019**, *352*, 79–90. [CrossRef]
14. Wang, Y.; Chang, Y.; Liu, M.; Zhang, A.; Guo, X. A facile strategy to prepare shaped ZSM-5 catalysts with enhanced para-xylene selectivity and stability for toluene methylation: The effect of in situ modification by attapulgite. *Molecules* **2019**, *24*, 3462. [CrossRef] [PubMed]

15. Hernando, H.; Ochoa-Fernandez, C.; Shamzhy, M.; Moreno, I.; Fermoso, J.; Pizarro, P.; Coronado, J.M.; Cejka, J.; Serrano, D.P. The crucial role of clay binders in the performance of ZSM-5 based materials for biomass catalytic pyrolysis. *Catal. Sci. Technol.* **2019**, *9*, 789–802. [CrossRef]
16. Delgado, J.A.; Águeda, V.I.; Uguina, M.A.; Brea, P.; Grande, C.A. Comparison and evaluation of agglomerated MOFs in biohydrogen purification by means of pressure swing adsorption (PSA). *Chem. Eng. J.* **2017**, *326*, 117–129. [CrossRef]
17. Zhang, H.P.; Li, W.; Xiao, W.D. Attrition Resistant Catalyst of Direct Dimethyl Ether Synthesis. *Appl. Mech. Mater.* **2012**, *174*, 97–101. [CrossRef]
18. Marigo, M.; Cairns, D.L.; Bowen, J.; Ingram, A.; Stitt, E.H. Relationship between single and bulk mechanical properties for zeolite ZSM5 spray-dried particles. *Particuology* **2014**, *14*, 130–138. [CrossRef]
19. Chaemchuen, S.; Zhou, K.; Mousavi, B.; Ghadamyari, M.; Heynderickx, P.M.; Zhuiykov, S.; Yusubov, M.S.; Verpoort, F. Spray drying of zeolitic imidazolate frameworks: Investigation of crystal formation and properties. *CrystEngComm* **2018**, *20*, 3601–3608. [CrossRef]
20. Spjelkavik, A.I.; Aarti Divekar, S.; Didriksen, T.; Blom, R. Forming MOFs into Spheres by Use of Molecular Gastronomy Methods. *Chem. Eur. J.* **2014**, *20*, 8973–8978. [CrossRef]
21. Lee, D.W.; Didriksen, R.; Olsbye, U.; Blom, R.; Grande, C.A. Shaping of metal-organic framework UiO-66 using alginates: Effect of operation variables. *Sep. Purif. Technol.* **2020**, *235*, 116182. [CrossRef]
22. Finsy, V.; Ma, L.; Alaerts, L.; De Vos, D.E.; Baron, G.V.; Denayer, J.F.M. Separation of CO2/CH4 Mixtures with the MIL-53(Al) Metal–Organic Framework. *Microporous Mesoporous Mater.* **2009**, *120*, 221–227. [CrossRef]
23. Peterson, G.W.; DeCoste, J.B.; Glover, T.G.; Huang, Y.; Jasuja, H.; Walton, K.S. Effects of pelletization pressure on the physical and chemical properties of the metal–organic frameworks Cu3(BTC)2 and UiO-66. *Microporous Mesoporous Mater.* **2013**, *179*, 48–53. [CrossRef]
24. Ribeiro, R.P.P.L.; Antunes, C.L.; Garate, A.U.; Portela, A.F.; Plaza, M.G.; Mota, J.P.B.; Esteves, I.A.A.C. Binderless shaped metal-organic framework particles: Impact on carbon dioxide adsorption. *Microporous Mesoporous Mater.* **2019**, *275*, 111–121. [CrossRef]
25. Nayak, N.; Vitorino, N.; Frade, J.R.; Kovalevsky, A.V.; Alves, V.D.; Crespo, J.G.; Portugal, C.A.M. Design of alumina monoliths by emulsion-gel casting: Understanding the monolith structure from a rheological approach. *Mater. Des.* **2018**, *157*, 119–129. [CrossRef]
26. Lefevere, J.; Protasova, L.; Mullens, S.; Meynen, V. 3D-printing of hierarchical porous ZSM-5: The importance of the binder system. *Mater. Des.* **2017**, *134*, 331–341. [CrossRef]
27. Lefevere, J.; Mullens, S.; Meynen, V. The impact of formulation and 3D-printing on the catalytic properties of ZSM-5 zeolite. *Chem. Eng. J.* **2018**, *349*, 260–268. [CrossRef]
28. Couck, S.; Cousin-Saint-Remi, J.; Van der Perre, S.; Baron, G.V.; Minas, C.; Ruch, P.; Denayer, J.F.M. 3D-printed SAPO-34 monoliths for gas separation. *Microporous Mesoporous Mater.* **2018**, *255*, 185–191. [CrossRef]
29. Lind, A.; Vistad, Ø.; Sunding, M.F.; Andreassen, K.A.; Cavka, J.H.; Grande, C.A. Multi-purpose structured catalysts designed and manufactured by 3D printing. *Mater. Des.* **2020**, *187*, 108377. [CrossRef]
30. Akhtar, F.; Andersson, L.; Ogunwumi, S.; Hedin, N.; Bergström, L. Structuring Adsorbents and Catalysts by Processing of Porous Powders. *J. Eur. Ceram. Soc.* **2014**, *34*, 1643–1666. [CrossRef]
31. DeLuca, J.P.; Campbell, L.E. Monolithic Catalyst Supports. In *Advanced Materials in Catalysis*; Burton, J.J., Garten, R.L., Eds.; Academic Press: New York, NY, USA, 1977; pp. 293–324.
32. Shams, K.; Mirmohammadi, S.J. Preparation of 5A zeolite monolith granular extrudates using kaolin: Investigation of the effect of binder on sieving/adsorption properties using a mixture of linear and branched paraffin hydrocarbons. *Microporous Mesoporous Mater.* **2007**, *106*, 268–277. [CrossRef]
33. Calleja, G.; Pau, J.; Calles, J.A. Pure and Multicomponent Adsorption Equilibrium of Carbon Dioxide, Ethylene, and Propane on ZSM-5 Zeolites with Different Si/Al Ratios, J. *Chem. Eng. Data* **1998**, *43*, 994–1003. [CrossRef]
34. Talesh, S.S.A.; Fatemi, S.; Hashemi, S.J.; Ghasemi, M. Effect of Si/Al Ratio on CO2—CH4 Adsorption and Selectivity in Synthesized SAPO-34. *Sep. Sci. Technol.* **2010**, *45*, 1295–1301. [CrossRef]
35. Sarti, E.; Chenet, T.; Pasti, L.; Cavazzini, A.; Rodeghero, E.; Martucci, A. Effect of Silica Alumina Ratio and Thermal Treatment of Beta Zeolites on the Adsorption of Toluene from Aqueous Solutions. *Minerals* **2017**, *7*, 22. [CrossRef]
36. Grande, C.A.; Cavenati, S.; Barcia, P.; Hammer, J.; Fritz, H.G.; Rodrigues, A.E. Adsorption of Propane and Propylene in Zeolite 4A Honeycomb Monolith. *Chem. Eng. Sci.* **2006**, *61*, 3053–3067. [CrossRef]

37. Ohlin, L.; Bazin, P.; Thibault-Starzyk, F.; Hedlund, J.; Grahn, M. Adsorption of CO2, CH4, and H2O in Zeolite ZSM-5 Studied Using In Situ ATR-FTIR Spectroscopy. *J. Phys. Chem. C* **2013**, *117*, 16972–16982. [CrossRef]
38. Grande, C.A.; Mondino, G.; Lind, A.; Vistad, Ø.; Akporiaye, D. Selective Removal of CH4 from CH4/CO/H2 Mixtures. In *Small-Scale Gas to Liquid Fuel Synthesis*; Kanellopoulos, N., Ed.; CRC Press: Boca Raton, FL, USA, 2015.
39. Heymans, N.; Alban, B.; Moreau, S.; De Weireld, G. Experimental and theoretical study of the adsorption of pure molecules and binary systems containing methane, carbon monoxide, carbon dioxide and nitrogen. Application to the syngas generation. *Chem. Eng. Sci.* **2011**, *66*, 3850–3858. [CrossRef]
40. Dunne, J.A.; Rao, M.; Sircar, S.; Gorte, R.J.; Myers, A.L. Calorimetric heats of adsorption and adsorption isotherms 2. O2, N2, Ar, CO2, CH4, C2H6 and SF6 on NaX, H-ZSM-5 and Na-ZSM-5 zeolites. *Langmuir* **1996**, *12*, 5896–5904. [CrossRef]
41. Mathias, P.M.; Kumar, R.; Moyer, J.D.; Schork, J.M.; Srinivasan, S.R.; Auvil, S.R.; Talu, O. Correlation of Multicomponent Gas Adsorption by the Dual-Site Langmuir Model. Application to Nitrogen/Oxygen Adsorption on 5A-Zeolite. *Ind. Eng. Chem. Res.* **1996**, *35*, 2477–2483. [CrossRef]
42. Olson, D.H.; Haag, W.O.; Borghard, W.S. Use of water as a probe of zeolitic properties: Interaction of water with HZSM-5. *Microporous Mesoporous Mater.* **2000**, *35*, 435–446. [CrossRef]
43. Bolis, V.; Busco, C.; Ugliengo, P. Thermodynamic Study of Water Adsorption in High-Silica Zeolites. *J. Phys. Chem. B* **2006**, *110*, 14849–14859. [CrossRef]

MDPI
St. Alban-Anlage 66
4052 Basel
Switzerland
Tel. +41 61 683 77 34
Fax +41 61 302 89 18
www.mdpi.com

Energies Editorial Office
E-mail: energies@mdpi.com
www.mdpi.com/journal/energies

www.ingramcontent.com/pod-product-compliance
Lightning Source LLC
Chambersburg PA
CBHW041954260926
51850CB00019B/2780